Make:

GETTING STARTED
WITH 3D PRINTING
2ND EDITION

Liza Wallach Kloski, Nick Kloski
HoneyPoint3D (TM)

Make: Getting Started with 3D Printing, 2nd Edition

By Liza Wallach Kloski, Nick Kloski

Published by Make: Community LLC

150 Todd Road, Suite 200, Santa Rosa, CA 95407

Make: books may be purchased for educational, business, or sales promotional use.
Online editions are also available for most titles.
For more information, contact our corporate/institutional
Sales department: 800-998-9938

Publisher: Dale Dougherty
Editor: Patrick Di Justo
Creative Director: Juliann Brown
Design: Jason Babler

This book's body font is Din Pro Light. Subheads are set in Geogrotesque Condensed.

June 2016: First Edition
Revision History for the Second Edition
May 18, 2021

See www.oreilly.com/catalog/errata.csp?isbn=9781680456431 for release details.

9781680456431

O'REILLY ONLINE LEARNING

For more than 40 years, www.oreilly.com has provided technology and business training, knowledge, and insight to help companies succeed.

Our unique network of experts and innovators share their knowledge and expertise through books, articles, conferences, and our online learning platform. O'Reilly's online learning platform gives you on-demand access to live training courses, in-depth learning paths, interactive coding environments, and a vast collection of text and video from O'Reilly and 200+ other publishers. For more information, please visit www.oreilly.com

HOW TO CONTACT US:

Please address comments and questions concerning this book to the publisher:

Make: Community LLC
150 Todd Road, Suite 200, Santa Rosa, CA 95407

You can also send comments and questions to us by email at books@make.co.

Make: Community is a growing, global association of makers who are shaping the future of education and democratizing innovation. Through *Make:* magazine, and 200+ annual Maker Faires, *Make:* books, and more, we share the know-how of makers and promote the practice of making in schools, libraries and homes.

To learn more about *Make:* visit us at make.co.

This book is lovingly dedicated to
the memory of Susan Diane Wallach.

A lifetime of teaching, artistry, and
inspiration.

A true maker.

PREFACE

A great deal has changed since the 2016 edition of this book. Where once we assumed that 3D printers were going to be everywhere, we now know that they will be in many places, but it may take some time before they'll wind up in every home. Still, if you're reading this book, that means that you're about to embark on a potentially life-changing journey.

Join the newest generation of makers as they wield the tools of digital fabrication to transmute thought into reality. The CNC machine carves metal, the laser cutter etches wood, and the PCB writer harnesses the power of electricity to drive even the most complex of robotics. But there may be no tool of alchemy more powerful than the 3D printer.

Follow along with this book, and you will be initiated into an elite society of alchemists who are capable of transforming the very matter of the physical world around you. Like me, you will become a maker.

I can tell you from personal experience that learning how to 3D print may not always be fun, but it will always be rewarding. Life, when you love 3D printing, can be a tumultuous one of clogged print heads, broken belt drives, and unruly 3D CAD models. There is hope, but you must start somewhere. And there's no better place to start than with this book, written by two educational alchemists—Liza Wallach Kloski and Nick Kloski of HoneyPoint3D™.

They have already empowered many thousands of apprentices, having successfully launched their own wildly popular educational program devoted to teaching the ins and outs of 3D printing. Liza and Nick will be able to accompany you on your journey as two Sherpa guides who will be there for you when your wires are crossed, your extruders are clogged, and your print bed isn't level.

Soon, you will understand how to turn the mental into the physical, the digital into the material. You'll be able to clone everyday items, teleport them across thousands of miles, or reconfigure them to suit your own needs. You'll befriend other mystics, like yourself, who can help you to hone your powers. You might even join a local chapter of the elders, a makerspace, where secret rites and rituals will see you unite your powers with the collective to unleash even mightier projects on the world at large. Or you could introduce the technology to your own business for the engineering of new designs, products, tools, jigs and fixtures.

That is what you can expect when you end your journey. Once you complete this guide, you, my friend, will become one of the initiated. You will be a maker.

—Michael Molitch-Hou
Editor-in-Chief
3DPrint.com
The Voice of 3D Printing/Additive Manufacturing

TABLE OF CONTENTS

FOREWORD: AN INTRODUCTION TO 3D PRINTING

"Feel fortunate! You are living at a time when technology is helping people become masters of their environments. A 3D printer puts the power of a manufacturing plant on your desk and opens worlds of opportunities that you (and the rest of humanity) have never experienced before."

That's what we wrote back in 2016 on the first page of the first edition. That year seemed like a pie-in-the-sky utopia of skyrocketing 3D printing stocks, new 3D business companies popping up overnight, an endless wave of 3D printing crowdfunding campaigns, and the promise of a 3D printer in every home.

Those things may still come to pass, but the media's interest in consumer 3D printing has waned while the usefulness of the technology in education and business has soared.

It is an adage that many new technologies follow a cycle of rapid interest ("hype") followed by the realization of what the technology can actually do (the "reality"). 3D Printing is in the Plateau of Productivity phase (**Figure F-1**) where real-world benefits are at an acceptable level of reliability and quality to be accepted by businesses for ongoing utilization (reduced risk of adoption). The prevalence of the technology accelerates rapidly as a result. The real advancements happen, perhaps without the fanfare, but with all the substance.

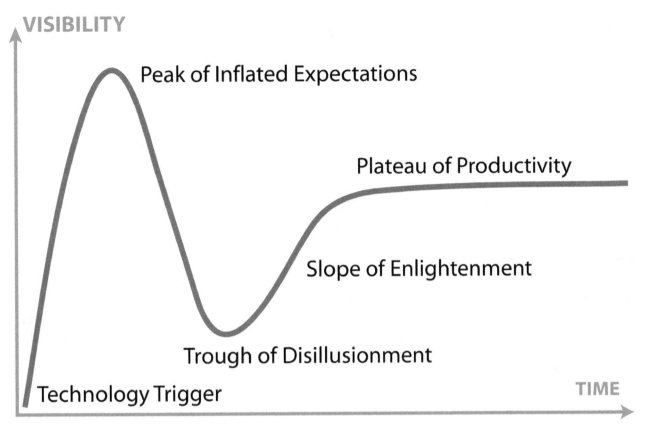

FIGURE F-1: Hype Cycle, credit Jeremy Kemp at English Wikipedia, CC BY-SA 3.0

The current examples of how 3D printing has and will continue to transform medicine, construction, education, and many other industries are the new stories. And educators have a special place in this moment: they are educating the future entrepreneurs, engineers, and researchers of tomorrow who will be using 3D printers to create viable human organs, colonize Mars, and who knows what else!

You will learn about the new 3D printing landscape that has emerged. One that has less hype, but more utility. One that focuses less on media attention and more on actual advancements in materials, printer technology, and use cases.

In this book we will explore the real-life use cases of 3D printing, how to manage a 3D printer yourself (with a visual troubleshooting guide), how you can incorporate 3D printing at home and the classroom, and a new chapter on how you can use 3D printing for your business.

A VERY, VERY BRIEF INTRO TO 3D PRINTING:

Industrial 3D printing has been around since the 1980s. The technology became available to hobbyists and consumers in 2009 when the RepRap Project brought together thinkers and coders from around the world to create freely open plans and software. This gave anyone the ability to build a personal 3D printer, such as the example shown in **Figure F-2**.

Fast-forwarding to today, the 3D printing industry (industrial and consumer combined) is now estimated at around $17.8 billion and is expected to be $23.9 billion in 2022 and $35.6 billion in 2024. In the first edition of this book, published in 2016, we cited the global 3D printing market was around $6-$7 billion; that's a lot of growth in 4 years! No one has a crystal ball to know the future, but it seems certain to us that 3D printing will increasingly affect the way we as a society design and manufacture physical items.

HOW 3D PRINTERS WORK

Now let's dive into the details. The most common type of 3D printing is known as "additive manufacturing" in which small amounts of material are slowly built up into an object. The primary method of this is called FDM (fused deposition modeling) or sometimes FFF (fused filament fabrication). FDM printers create objects by building up material layer by layer over time. A thin strand of filament feeds into a part of the machine called an extruder, which melts the plastic at a high temperature—typically around 200°C. There are many other methods of 3D printing that this book will cover in later chapters, and they all work in the same "additive" way. They bind, melt, or photo-polymerize (liquid resin which is hardened with UV light) material together to create physical geometry.

The FDM process is very similar to using a hot glue gun. You probably have one at home and have used it for craft or school projects. When you squeeze the handle of a hot glue gun, the solid glue is pressed against a heating element and soft, spaghetti-like strands of molten glue extrude from the nozzle. Imagine swirling those strands around and around on top of each other, forming circles that build up into a tube. As the glue cools, it hardens, and a tube is created.

Following the glue gun analogy, in a 3D printer a thin strand of melted plastic is programmed to deposit down, layer by layer, on a flat surface known as the build plate, where it cools and hardens into an object. The 3D printer knows precisely where to trace each layer through instructions in a digital file sent from a computer. The image in **Figure F-3** provides a closer look at how this technology works.

Most consumer 3D printers use FDM technology and have a large spool of coiled plastic called filament attached to them.

FIGURE F-2: An early example of a person assembling a 3D printer from a kit.

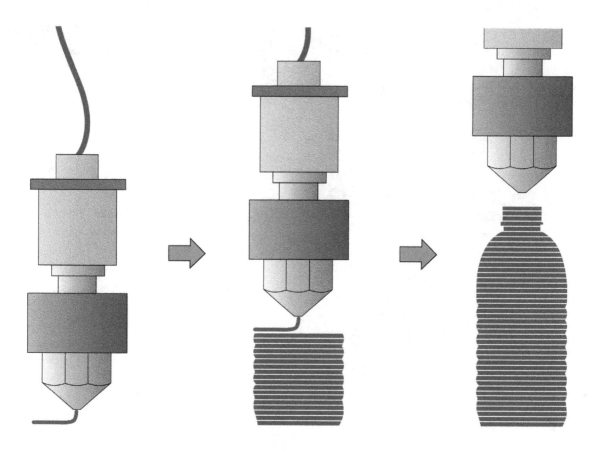

The extruder pushes the filament through the hot end of the nozzle which then melts the filament into very thin layers.

As the material exits the nozzle, it cools and hardens onto the build plate. Subsequent layers then bond to the previous layer below.

Over time, the 3D form will begin to take shape. Once the last layer is deposited, the 3D print is ready to be removed from the build plate.

FIGURE F-3: An extruder is depositing melted plastic one layer at a time to build a bottle (infographic HoneyPoint3D™)

3D PRINTING IS NOT LIKE 2D PRINTING

In 2013, at our then retail 3D printing store in Oakland, California, the majority of the public had not made the connection that 3D printers (which print three-dimensional objects) are very different from 2D printers (which print flat images on paper). We used to get a lot of questions from casual visitors along the lines of, "Where do you put the ink?" and "How much paper does it need?"

We would teach that you can't print a "flat" graphical image such as a PDF or JPEG in 3D. A special kind of digital file is needed to make a 3D print. That type of file is called a 3D model and it has three-dimensional information about the object you want to build.

Luckily for us, between 2009 and 2017, 3D printing received a tremendous amount of media attention, and the general public made huge leaps in awareness of the technology. 3D printing was a paradigm shift from the traditional understanding of 2D printing based on paper-and-ink printers to the 3D printers that now affect everyday life. We will go into more details about 3D printers and 3D models later in this book.

MANAGING EXPECTATIONS AS YOU START YOUR JOURNEY

Managing a 3D printer is more difficult than it should be. The technology is still relatively young, and the machines are not as plug-and-play as most home appliances, like a microwave oven. Creating a 3D file in order to 3D print something requires practice and skill. Expect to go through a learning curve where good old trial-and-error will give you lots of valuable experience and, eventually, success! You will learn the tricks of the trade as you go. Here are some common experiences you should expect:

• 3D PRINTS FAIL

A lot. Especially when you are first learning to use your 3D printer. Did you set the wrong temperature? Did the 3D file have errors in it? Was the print bed not level? You'll be asking yourself these questions and more as you experiment.

• 3D PRINTS TAKE A LONG TIME

Want to print a phone case? No problem! But the print will likely take 6+ hours.

• 3D PRINTERS NEED ONGOING MAINTENANCE

Motor belts pop off. The hot end gets jammed. Stuff happens and, in most cases, it will be you fixing it.

• SOMETIMES 3D PRINTS NEED PRE/POST-PROCESSING

Some 3D model files you find on the Internet can't be printed as-is and will need to be fixed. You will learn more about how to fix them in future chapters. 3D printed objects will need work to get the smooth surface you may desire. You will find that sandpaper does wonders with those rough edges on a newly hatched 3D print. (Hint: Grit 100 to 600, Medium, fine, and extra-fine)

DON'T WORRY! THIS IS WHY WE UPDATED THIS BOOK!

We know 3D printing isn't just a technology; it's an ecosystem of software, hardware, education, communities, and materials. Implementing 3D printing into your life takes a three-pronged approach: education, CAD file generation, and the physical aspect of 3D printing. The rewards of learning how to 3D print are well worth the learning curve, and you'll be glad you made the effort. Reading this book will save you time, money, headaches, and heartaches and will lessen the number of "valuable lessons" you have to go through on your own. We made plenty of mistakes so you won't have to!

WHAT'S IN THIS BOOK?

Getting Started with 3D Printing 2nd Edition offers you a clear, updated roadmap of best practices to help you successfully bring 3D printing into your home, classroom, and business. We hope you will find it a fun, practical guide that will help you navigate your way from initial curiosity to active, hands-on 3D printing. The book is intended for those who have no or little prior experience in 3D printing. If you do have experience with 3D printing, this book will help you troubleshoot your failed prints, expand your industry knowledge, and offer you tips on making 3D models.

THIS BOOK WILL INFORM YOU ABOUT:

- How 3D printers work
- How people are using 3D printing in the classroom, home, and business
- The different 3D printing Technologies
- What to look for when buying a 3D printer and supplies
- Setting up and maintaining your own 3D printer
- Outsourcing 3D modeling and printing services
- The workflow from idea to 3D print
- Creating and fixing your own 3D models
- Designing a personal 3D printing makerspace
- How to make money with 3D printing
- The future of 3D printing

Fasten your seatbelt and prepare yourself for this fast approaching "manufacturing revolution."

It may just change your life!

"3D Printing is revolutionizing how we make and use everyday objects. You need to get on board now, and this book is a great place to start!"

–NORA TOURE
Founder; Women in 3D Printing

PART I: APPLICATIONS OF 3D PRINTING

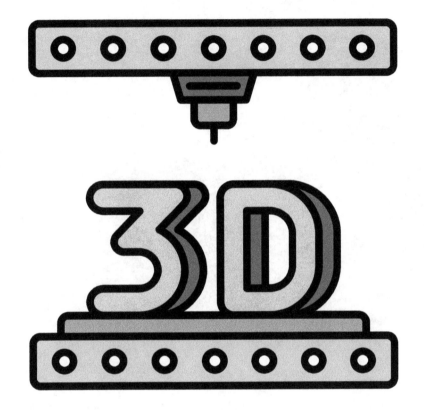

CHAPTER **1**

INTRODUCTION

FIGURE 1-1: Workers assembling condensers at the Atwater Kent Factory in North Philadelphia, 1925.

THERE IS NO DOUBT that the rise of 3D printing laid the foundation for a significant change in our society in the coming decades. This chapter gives you a short historical perspective on the manufacturing roots of 3D printing and how it currently affects our lives.

YOU SAY YOU WANT A REVOLUTION?

The adoption of 3D printing is being heralded by some as the beginning of a "third industrial revolution," but we feel it is better called the "personal manufacturing revolution."

The first industrial revolution developed new manufacturing processes that included going from hand production methods to machines, old to new chemical manufacturing processes, and the creation of machine tools. The second industrial revolution is a continuation of the first industrial revolution and was characterized with the increased adoption of the steam transport, large-scale manufacture of machine tools, the increase in the use of steam-powered machines, and mass production **(Figure 1-1)**. The second industrial revolution, which introduced assembly lines, continues to affect how we manufacture goods today. These two revolutions marked major turning points in history; almost every aspect of daily life was influenced in some way. Average incomes and population numbers showed unprecedented growth, and the standard of living rose for most people.

3D printing, like its fabrication predecessors, promises to change not just manufacturing but our way of life. The key difference is that 3D printing gives power to the individual. Essentially, it is a factory on your desk; you can model an idea and 3D print the object the same day. You can manifest your concepts into a physical form which was only once available through expensive, exclusive, and time-consuming industrial prototyping. You don't need permission from a board of directors, or even orders from customers, to produce new products. You just need your imagination and a spool of plastic.

This is why we and many others believe "personal manufacturing revolution" is a more fitting term for the changes 3D printing is bringing. It shifts the focus and ability to the individual, showcasing self-expression and self-sufficiency.

A MANUFACTURING FULL CIRCLE

3D printing won't completely replace traditional manufacturing processes. Rather, it will augment the current means of mass production. In addition,

- Scalable workforce
- Production of one, a few OR many
- Not location dependent
- Even higher standard of living
- Quick, cheap, and custom

- Single worker production
- Low production volume
- Rural manufacturing
- Low standard of living
- Slow, expensive, but custom

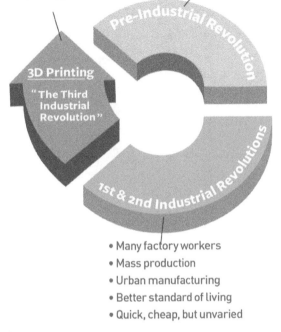

- Many factory workers
- Mass production
- Urban manufacturing
- Better standard of living
- Quick, cheap, but unvaried

FIGURE 1-2: 3D printing combines the best of all manufacturing periods and shows how manufacturing has come full circle (infographic by HoneyPoint3D™)

the old centers of manufacturing (large factories in urban areas) may no longer be as necessary. You can now manufacture locally in rural areas, both in large factories and small shops, and you can economically produce one piece or many. In this way, 3D printing has combined the best aspects of pre- and post-revolution manufacturing, as shown in **Figure 1-2.**

Local design and production can be quick, efficient, and environmentally conscious. And because each 3D print is made individually, modifications can be added quickly between 3D prints (something that mass manufacturing cannot do easily).

In the next chapter, we'll examine how 3D printing affected the consumer maker movement.

2

3D PRINTING AND THE MAKER MOVEMENT

FIGURE 2-1: Nick Kloski making a presentation on 3D printing at the East Bay Mini Maker Faire, Oakland, California

IF YOU'RE READING this book, you probably find the act of creating very satisfying. 3D printing is a relatively new, inexpensive, and readily available technology that allows makers to do things they couldn't do before. Inexpensive 3D printers are enabling ordinary people to make their own jewelry, toys, mechanical parts, and many other items. 3D Printing is also a technology that all ages can participate in. Students as young as 7 years old are learning how to CAD model and see their creations come to life. Venues for learning about this technology range from maker camps, library-sponsored classes, to paid online courses and live seminars.

IMPACT OF THE MAKER MOVEMENT

The "maker movement," can be seen as an extension of the "DIY" (do-it-yourself) community and, from the start, 3D printing was a natural interest for people that liked taking their own creativity into their own hands. The maker movement is still an extension of the DIY mindset and has had a tremendous influence in promoting 3D printing through the hundreds of thousands of people attending the various Maker Faires across the world. **Figure 2-1** shows attendees at a local Mini Maker Faire listening to our lecture on 3D printing. No matter what experience level you have, you can be a maker, too. And, as you will discover in this chapter, a whole ecosystem of companies, organizations, and services are available to help you use 3D printing to foster that ability in you. Having the attitude of a maker will help you imagine and trial and error your way into bringing 3D printing into your school, home, or business.

You can see where the next faire is located by searching for "Maker Faire featured faires" in your search engine of choice. These faires were critical to the adoption of customer 3D printing, which then paved the way for businesses to use 3D printed parts with consumer acceptance.

WHO IS A MAKER?

Each year many people attend Maker Faires across the globe. So, who are these members of the maker community?

A maker is anyone who puts things together creatively. Look around you, makers are everywhere. The do-it-yourselfer making a wood door for his house is a maker. The hobbyist assembling a flying drone from a kit is a maker. A programmer developing an Arduino-based electronic device is a maker. The educator creating a unique learning tool is a maker. They are all making something new by using their hands and creativity, turning their ideas into physical forms.

In fact, you probably already are a maker in your daily life! Maybe something broke in your kitchen and you hacked a solution together to make it work again...if so, you're a maker! The children in schools making their 7th-grade science project...they are makers too! You don't have to have a garage full of tools to be a maker, you just need to think and build creatively. 3D printing can turn your ideas to form.

As you tinker and experiment with this technology, you will undoubtedly find ways to use 3D printing to enhance your life. In fact, a study from Michigan Technological University estimated that a family can save anywhere from $300 to almost $2,000 a year by 3D printing items such as combs, cookie cutters, door stops, tool parts, and more. It's easy to see how 3D printing has found a happy home in the maker movement.

HOW THE 3D PRINTING ECOSYSTEM HELPS YOU BE A MAKER

An entire ecosystem of products and services has developed around this technology. Whether you are a maker, investor, educator, business startup, or just testing the waters, it's important to be aware of the companies and services that make up the current 3D printing ecosystem.

The number of related goods and services offering support in the 3D printing landscape has grown tremendously since we wrote the first edition of this book. Following are four charts that highlight valuable referrals in the 3D printing industry. It's important to note this reference list of companies and organizations is not comprehensive, but rather an overview of what you can expect to encounter in the 3D printing world.

You'll see you don't have to master all aspects of 3D printing yourself; you can download ready-made 3D models, hire 3D design and CAD services, outsource the 3D printing to a service bureau, or even purchase 3D printed objects. How you want to approach it is up to you!

In the next chapter, we'll examine some of the ways that 3D printing is being used today to provide greater choice in consumer products, better health care, and more. 3D printing is already having an important impact on our society.

LEARNING

3D printing visual aides to supplement learning. 3D printing-centric course curriculum. Creation of after-school fab-labs.

MAKER-SPIRITED

Toys for children. A way for parents to engage with children in a maker-spirited environment.

HOME COOKING

Custom cookie cutters, ice cube molds, and other household kitchen items.

How everyday Makers use 3D printing

HOBBIES

3D printing their own custom pieces for drone kits and various remote-controlled vehicles and gadgets.

HOME REPAIRS

Replacement parts for appliances and other objects around the house; for example, outlet covers, pieces for washer/dryer, doorstops, wall hooks.

MEMORY KEEPSAKES

3D scanning and printing miniature self, family portraits, wedding cake toppers.

FIGURE 2-2: Examples of how consumers, turned makers, are using 3D printing today. Infographic created by HoneyPoint3D™

3D Printer Manufacturers

There are over 150 consumer 3D printers on the market costing anywhere from $200 to $5000. Available as kits, or as fully assembled units, each printer has different specifications that make it unique (e.g., maximum build volume, types of materials you can print, maximum temperature, etc.)

Kudo3D, FSL3D, Airwolf3D, Formlabs, Ultimaker Afinia, RepRap, 3D Systems, Stratasys ZMorph, MakerBot, XYZ Printing, Gigabot, Raise3D, Prusa Research, SeeMeCNC

Material Suppliers

All 3D printers require a material to print with, and you have a variety of materials to choose from. The printer hardware will determine what materials can be printed. Here is a list of popular material suppliers.

Taulman3D, Maker Juice, ColorFabb, Reprap Austria, Asiga, MatterHackers, Proto-pasta, ESUN 3D, Faberdashery, Siraya Tech

3D Software

In the realm of 3D printing, software is vital. From slicers to 3D modeling tools, you will be interacting with a piece of software at some point during your 3D printing experience. We have listed some great programs below, many of which are free!

Fusion 360, Tinkercad, SolidWorks, Blender, Meshmixer, AutoCAD, Inventor, SketchUp, 3DCoat, Moment of Inspiration, Open SCAD

FIGURE 2-3: Chart showing examples of manufacturers, suppliers and software.

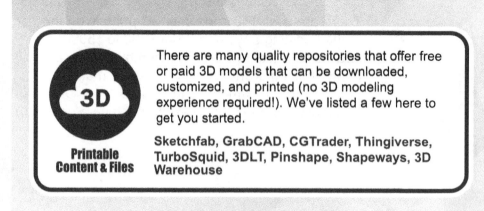

There are many quality repositories that offer free or paid 3D models that can be downloaded, customized, and printed (no 3D modeling experience required!). We've listed a few here to get you started.

Printable Content & Files

Sketchfab, GrabCAD, CGTrader, Thingiverse, TurboSquid, 3DLT, Pinshape, Shapeways, 3D Warehouse

You don't need a 3D printer to 3D print! There are many companies that offer professional printing services as well as design services to help you bridge the gap from idea to physical form. This is a great option if you wish to have more materials and size options at your disposal.

Outsourced Print & Design Services

Sculpteo, Shapeways, Protolabs, 3D Hubs, HoneyPoint3D, 3D Systems, FATHOM MFG, RedEye, Stratasys Direct

If you need a 3D model of an already existing object, 3D scanning is a very effective method. You can create scans for free using a simple digital camera, or there is a wide variety of professional solutions available.

3D Scanning

Artec3D, MakerBot, FARO, David-3D (HP), Scansite, Autodesk ReCap, HoneyPoint3D, OpenScan

FIGURE 2-4: Chart detailing places to find models online, where to get those 3D files printed, and ways to get physical objects into a 3D file using 3D scanning

Media & Education

There are various educational resources that teach applicable skills for students and professionals who want to learn more about 3D printing. Furthermore, there are countless books, newsletters, magazines, and forums that will help you stay informed of new trends and developments in the industry. We've listed some popular sources below.

Make Magazine, Make.co, 3DPI.com, 3DPrint.com, 3Ders.org, TCT Magazine, Print Shift, HoneyPoint3D

Finished Products

There are some companies that sell consumer-ready 3D printed products that you can purchase directly. This approach follows the traditional shopping experience where customizability is limited but the design is high quality.

Robohand, Amazon, Shapeways, Nervous System, 3DLT

Retail Shops

3D printing retail shops are a relatively new concept, but are becoming more common as the technology advances. They offer a great physical location where you can buy 3D printers or 3D printing services.

iGo3D, iMakr, The UPS Stores, The 3D Printing Store

FIGURE 2-5: Chart detailing consumer-focused 3D printing events, learning venues, and research organizations.

Industry Event Organizers

Events are a great way to stay informed on the latest developments in the industry. They also offer an opportunity to see 3D printing technologies in person, and in most cases meet the people developing them. These events happen all throughout the year, so check out these organizers to find the next one happening near you!

Rising Media, Make.co, 3D Printshow, TCT Magazine, The 3D Printing Association, Consumer Electronics Show (CES), FabCon 3.D, IDTechEx 3D Printing, LinkedIn, Meetup, RAPID

Community Maker Spaces

If you're looking for a place to go tinker, then a fab lab or hackerspace is for you. Although they carry other Maker equipment like CNC machines and laser cutters, you'll likely find a 3D printer that you can experiment with. They offer friendly, community-minded environments for learning.

The Fab Lab, FabCafes, MakersFactory, Deezmaker, many local libraries, a Community Makerspace near you!

Legal & Market Research

3D printing technology continues to develop. As such, there are many legal and economic issues that come along with it. Here are some companies that are actively researching 3D printing and its worldwide impact.

Wohlers Report, SMARTech, Finnegan, Gartner, 3DPI.com, America Makes, 3D Hubs, Senvol

FIGURE 2-6: Some of the organizations, companies and resources that shape the 3D printing ecosystem. (All infographics by HoneyPoint3D™)

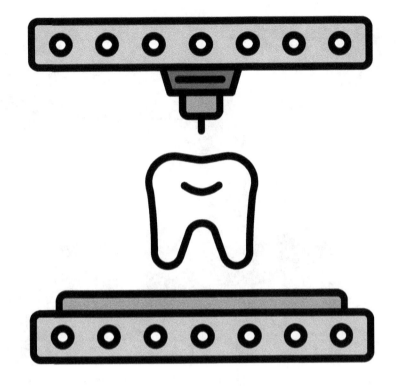

HOW 3D PRINTING IS BEING USED TODAY

Photo Courtesy Carson Chan

FIGURE 3-1: Prototyping process for a Mazda collector car. HoneyPoint3D's CAD model of rare hubcap, based solely on a photo (below). 3D printed / CNC milled hubcap on award-winning car (above).

MAKERS ARE NOT the only ones using this technology to further advancements in every area of life. Companies and other organizations have adopted elements of 3D printing to enhance, improve, and even create their products and services that are used in everyday life.

In this chapter, we'll look briefly at some of the innovative ways individuals and organizations are currently using 3D printing technology and why it's gaining more momentum.

RAPID PROTOTYPING FOR YOUR IDEAS, DESIGNS, AND INVENTIONS

Have you ever wanted to drive a car that had a steering wheel made just for you? Or how about door handles that are based on actual 3D scans of the grip pattern of someone with rheumatoid arthritis, to allow them to open doors at home more effectively? You can have it made with 3D printing. In fact, **Figure 3-1** shows that you can even replace life-size hubcaps for a collectible car using 3D printing.

Maybe you don't need to replace the hubcaps on a rare sports car, but what about a personalized 3D printed luggage tag or a replacement part no longer in production? Rapid prototyping is defined as the ability to quickly fabricate a model of a physical part using three-dimensional computer-aided design (CAD) soft-

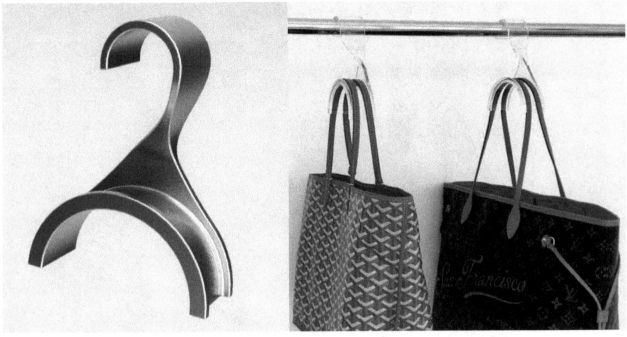

FIGURE 3-2: Luxe Bag Care hanger in CAD form (left) and as the actual product made from clear acrylic (right)

ware. 3D printing is ideal for rapid prototyping in that you can make changes quickly and produce a sample of one, saving time and money.

In the rapid prototyping division of our company, we see hundreds of people just like you looking at ways to invent and personalize the objects around them. They often come to us with hand drawings, a physical prototype, mechanical drawings, or even just ideas! We then translate this information into a CAD model (i.e., a digital file). The first 3D print lets our clients test the design. If adjustments are needed, we modify the CAD file and then 3D print the next version. This process continues until the client is satisfied with the final version of their product.

Figure 3-2 shows an example of how one of our clients turned an idea she had into a product sold at The Container Store. Individuals, small businesses, and large corporations everywhere can now customize products in small production runs.

Maybe you have an invention? Think of your possibilities! Instead of expensive tooling costs that can run into the tens of thousands of dollars, you can now test your ideas with a 3D printer and, in many cases, get your first prototype for under $50 USD per 3D print. And because changing an aspect of the design in a CAD modeling program is easy, each and every object can be subtly different for each buyer. This is something injection molding either cannot do or would require extensive postprocessing in order to achieve.

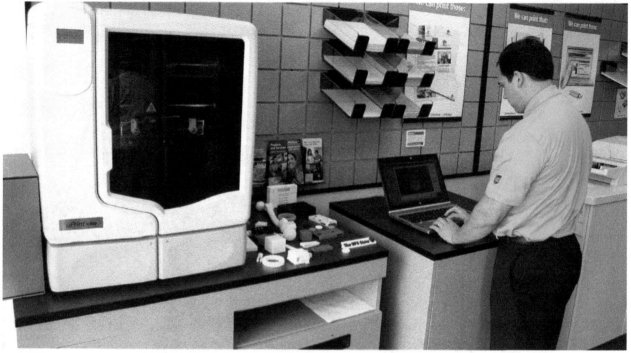

FIGURE 3-3: The UPS Store currently has 34 physical locations throughout the United States that offer 3D printing services (photo credit: The UPS Store)

Never before has there been such a versatile manufacturing tool that is so accessible to the masses. Your rapid prototype could be an iteration of a design or the final product. It's up to you!

FORWARD-THINKING COMPANIES ARE ACTIVELY BRINGING YOU 3D PRINTING

In 2013, The UPS Store, the world's largest franchisor of retail shipping, postal, printing, and business service centers, started offering local in-store 3D printing services using the Stratasys uPrint SE Plus 3D printer. Now, select The UPS Store locations across the country offer potential "same day" 3D printing services to their customers (**Figure 3-3**). These offerings brought retail 3D printing to a national level in the United States.

DON'T WORRY, COMPLEXITY WON'T COST YOU MORE

Do companies charge more for more complex products? When you buy something, it's almost always the case that the more complex or ornate versions cost more. This is not true with 3D printing. You don't need to pay more for the highly complicated, complex designs, other than the time to design them yourself. The 3D printer just "reads" where to put the material...it doesn't "care" if it's simple or complex. The 3D printer just registers how much material is being used, not how intricate the design is, so your cost is based on the material used and not detail.

This is another reason why 3D printing is important. Companies (and you!) can now design highly complex objects (**Figure 3-4**, **Figure 3-5**, **Figure 3-6**, **Figure 3-7**) that were once either impossible or too expensive to make with traditional manufacturing techniques like injection molding.

The objects shown in **Figure 3-8** and **Figure 3-9** are also great examples of 3D printing's capabilities. **Figure 3-8** shows how 3D printing allowed the industrial designer to create very organic shapes that are also precise and **Figure 3-9** shows how an artist included inner layers that were created during the printing process.

FIGURE 3-4: Images of 3D printed stop-motion figures used by Oscar nominee Brian McLean in the characters, props, and sets of his four LAIKA films (Kubo and the Two Strings, The Boxtrolls, Paranorman, and Coraline). Photos courtesy of FathomMFG, Oakland, CA

FIG 3-6: 4th Annual International Conference and Exhibition of 3D printing and scanning in Moscow, Russia. Examples of people 3D printed from files created using 3D scanning. Production of 1 unit is as economical as large quantities. The scan files can be resized and printed in different sizes.

FIG 3-9: A complex 3D printed artwork that would be impossible to produce using common manufacturing techniques such as injection molding.

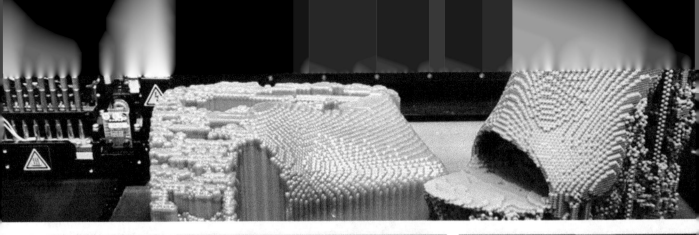

FIG 3-7: Designer Francis Bitonti and FathomMFG partnered together to create a 3D printed "Molecule Shoe"

FIG 3-8: Geneva Motor Show, EDAG 3D printed Soulmate Concept car, front-side view. Geneva, Switzerland.

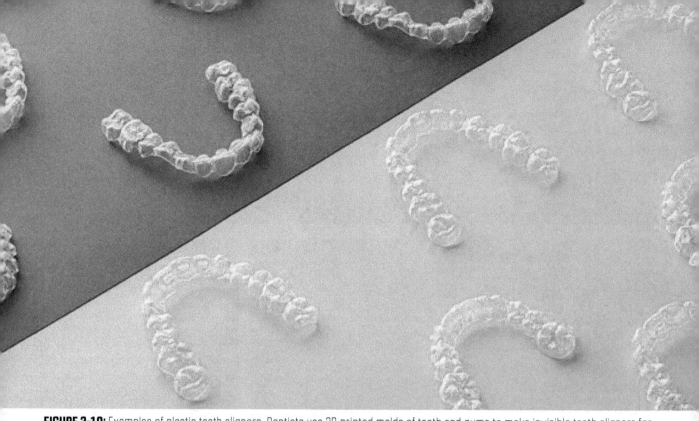

FIGURE 3-10: Examples of plastic teeth aligners. Dentists use 3D printed molds of teeth and gums to make invisible teeth aligners for patients

3D PRINTING IS ADVANCING HEALTH CARE: CUSTOMIZED MEDICAL APPLIANCES

Have you heard of a company called Smile Direct? Smile Direct manufactures clear dental retainers that are perfectly formed to the shape of your own teeth, guiding them to be more straight over time. Smile Direct sends you a dental mold kit you bite into and send back. They then make a 3D printed mold of your teeth. From that mold, they are able to make the retainers around that form. Each retainer is customized and no two are exactly the same as shown in this concept image (**Figure 3-10**). Smile Direct makes millions of sets of retainers per year and because they rely on the accuracy of digital manufacturing, customers do not need to see a dentist in person.

Another example of how the dental industry is using 3D printing can be illustrated by how ceramic caps can be made for patients that want to change the shape of their teeth. A dentist will create a mold of a patient's teeth and 3D print the mold using a detailed high resolution resin printer. The ceramic veneers are designed based on the digital 3D printed model as shown in **Figure 3-11**.

Another example of what we feel could be a very large application in medical appliances is customiz-

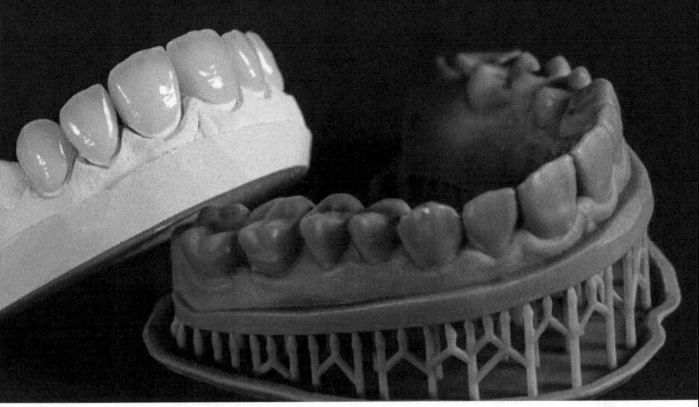

FIGURE 3-11: A patient's teeth 3D printed using high detail grey resin and an analog model cast from white gypsum (the traditional way of making dental models). The veneers on the gypsum mold were fitted using the 3D printed models.

able casts for bone injuries. **Figure 3-12** shows the concept of 3D scanning a part of a patient's anatomy, creating a 3D printed cast based on that particular patient's geometry, and then being able to adjust the model over time if needed.

New medical applications using 3D printing are being developed by other companies for hip replacements, prosthetics, and more. Not only can you use 3D printing in your life, but it can be a part of you, literally!

FIGURE 3-12: The process of creating a cast for a hand injury using 3D scanning, CAD modeling and 3D printing.

3D PRINTING HAS A SOLE

We can all relate to the sore feeling our feet have after walking around all day. Some people have more specialized needs for their feet due to medical

FIGURE 3-13: A custom-fitted 3D printed orthotic. The holes are for aesthetics as well as weight reduction and make those areas slightly more flexible, as dictated by the needs of the patient

conditions. The answer to alleviating that pain is often custom orthotics, but they are expensive and take some time to produce. Companies have developed processes which can take your exact foot and condition and create customized insoles using 3D printing. Rather than a lengthy fitting and hand fabrication process to receive custom orthotics, a consumer can use the camera on their cell phone (via a downloaded application), and a 3D model of the foot is generated just from those pictures. The 3D model is then used to fabricate a custom insole that is lighter, cheaper, and manufactured more quickly than one produced using traditional techniques (**Figure 3-13**). These 3D printed versions reduce the amount of fatigue both on your feet and wallet by providing 3D printed insoles at a fraction of the cost and length of time to produce.

3D printing is ushering in an era of personal mass customization. From custom orthotics to custom orthopedics, this technology can create hundreds of thousands of unique items for customers with all of the complex work being taken care of through the software. And the manufacturing is done on-demand, right when you order it! The capabilities demonstrated by these examples illustrate why 3D printing is becoming increasingly relevant in your day-to-day life.

YOU CAN GO BIG

When consumers think about 3D printing, they think about the $500 crowdfunded printers but 3D printing actually started with large, very expensive enterprise-level 3D printers that can cost up to $500,000 each. **Figures 3-14, 3-15, 3-16** and **3-17** show examples of these printers and what they can do.

FIGURE 3-14: A row of HP Multi Jet Fusion 3D printers. (Courtesy Forecast3D)

FIGURE 3-15: Huge print volume from a Stratasys Fortus 900mc printer (Courtesy Forecast3D)

Need to build a house? Why not use 3D printing to build some or all the walls? Believe it or not, companies are building large scale 3D printers that use cement to create parts or all of structures. The cement is laid down layer by layer as shown in **Figure 3-16**.

FIGURE 3-16: Rome, Italy. A huge industrial 3D printer builds a building made of cement automatically without the help of workers. The printer is guided by the computer program.

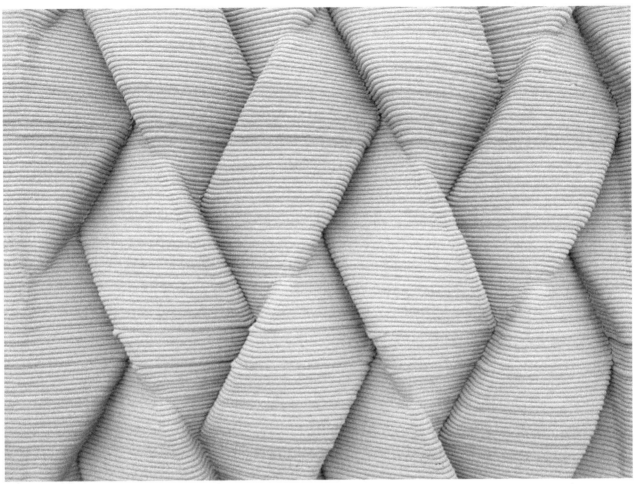

FIGURE 3-17: An example of a more refined 3D printed cement wall used in a housing project.

In the next chapter, we'll dive into the hands-on part of this book by describing the most common consumer 3D printing technology: fused deposition modeling, or FDM for short. Turn the page to learn more!

PART II: HARDWARE AND PRINTING CHOICES

CHAPTER

4

UNDERSTANDING FDM PRINTERS

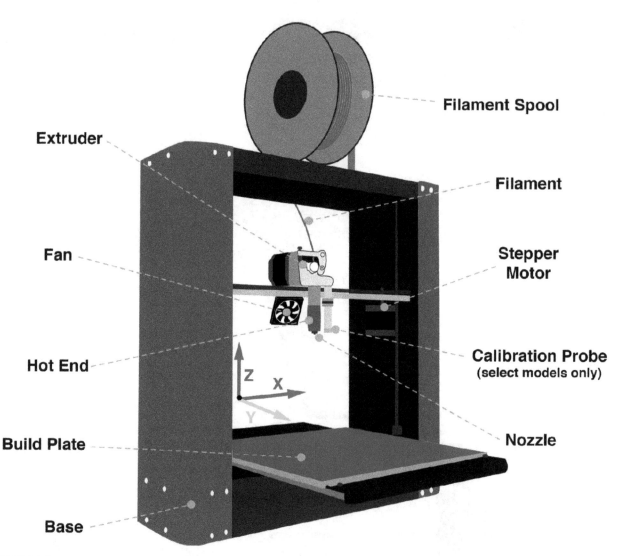

Filament Spool

Extruder

Filament

Fan

Stepper Motor

Hot End

Calibration Probe
(select models only)

Build Plate

Nozzle

Base

FIGURE 4-1: Anatomy of an FDM 3D printer (infographic by HoneyPoint3D™)

Consumer 3D printers are described by the material and process they use to create physical objects. There are two main descriptors of consumer 3D printers: those that use filament, and those that use liquid resin. The filament-based 3D printers are called "Fused Deposition Modeling" (FDM) machines, and the 3D printers which use light-hardened liquid resin are generically called "Resin Printers." In this chapter, we will focus on the FDM technology, which is the most commonly used consumer-level 3D printer. In the following chapter, we will focus on resin printers.

A fun fact: Many (if not all) of the consumer 3D printers being used today originated from a community-driven project called the "RepRap Project." This project started in 2005 in England and is open-sourced which means any future software development made must be available to all to use. This grass-roots community advanced FDM printing in the years to follow and the concept of keeping the advancements available to all was contested in the years to follow with some companies wanting to "close-source" their developments.

RepRap stands for "Replicating Rapid Prototyper." You can learn more about it at **http://RepRap.org**. This "Project" is a global community of tech-savvy people that came together to work on 3D printer hardware and software in an effort to create and refine freely available 3D printer designs, democratizing user access to this world-changing technology.

The first RepRap Project focused on the FDM printer, as FDM printing is more simple and straightforward than resin printing. In 2009, patents on FDM printing expired, allowing the world to delve into custom printer creation. They donated their time and work to the world, and to you.

Today, you can purchase a 3D printer kit you can assemble yourself by following detailed instructions, or you can buy a fully assembled unit that is designed to be usable right out of the box...the choice is up to you!

ALL FDM 3D PRINTERS WILL HAVE THE FOLLOWING COMPONENTS SHOWN IN FIGURE 4-1:

- ✅ Filament
- ✅ Extruder or extruder assembly
- ✅ Build plate/build area
- ✅ Linear movement components
- ✅ Frame / chassis
- ✅ Controller unit
- ✅ Stepper motor for the extruder assembly, to allow the build plate to move in one or more directions, and a motor to raise/lower the build plate

HERE'S A BASIC DESCRIPTION OF WHAT THE PARTS OF AN FDM 3D PRINTER DO:

- The BUILD PLATE (or PRINT BED) is a level, flat area where the 3D print starts.
- A FILAMENT SPOOL holds the plastic that will be melted.
- THIN FILAMENT (usually a type of thermoplastic) is wrapped around a filament spool and is the raw material the 3D printer uses to make objects. Consumer printers print with a filament that is one of two diameters: 1.75mm or 3mm. Diameters are not interchangeable, so be sure to research what your printer uses.
- The EXTRUDER is the name for the assembly that grabs the filament and pushes it down through the hot end (described below).

- STEPPER MOTORS control the belts to create the movement of the build plate and extruder.
- The CHASSIS is the frame of the 3D printer, which could be made from metal, plywood, plastic, etc.
- The HOT END is the part of the extruder that heats the filament to just the right temperature based on that material, and has a nozzle at the bottom to allow the molten filament to flow through.
- A FAN blows air over the in-process part, helping to remove excess heat from the physical model, making the depositing and bonding of the layers more successful.

TAKE NOTE: There are many consumer 3D printers on the market, with a variety of features and functions. For instance, some have print beds that do all the moving, while others rely on the extruder assembly to do all the movement. All are valid designs and particular choices come down to personal preference.

RESOLUTION LEVELS

Let's take a moment to talk about how 3D print quality is defined. In the more prevalent 2D printing world you will see printers that claim "600 dpi" or "1200 dpi" resolution. Those metrics refer to "dots per inch" of ink deposited on the paper. The more little dots of ink per inch, the higher the resolution. It's similar in 3D printing, but the process is about layer height and nozzle size!

ALL 3D PRINTERS (FDM AND RESIN) HAVE TWO DIFFERENT TYPES OF RESOLUTIONS:

- Z is the height resolution (in the up/down, or Z plane)
- XY is the positional resolution (in the left-right, forward-back, XY planes)

The Z-height resolution is the most common metric you will see in FDM printers. As you can see in **Figure 4-2**, these three cubes are all the same size, but they are 3D printed at different layer resolutions. All of the measurements given here are in "microns," which are fractions of a millimeter. 1,000 microns equal 1 mil-

FIGURE 4-2: A close-up image of three-layer heights printed at (left to right) 100 microns, 200 microns, and 300 microns

limeter, so we are talking about resolutions that are typically around 1/5th of a millimeter (or 200 microns).

In the three example prints in **Figure 4-2**, the individual layers are more densely packed on the left than on the middle or right. If you want a higher-quality surface finish, then you will choose a layer height that is smaller, which packs more layers into your end object. The lower the micron level, the less you can see the print lines on the object. 3D printing comes under scrutiny when people expect plastic 3D prints to look like an injection molded part, but they are different technologies with advantages and disadvantages. When we discuss resin 3D printing in the next chapter, you will see that the surface finish will look more like the plastic parts you would see at a store. This is something to consider when choosing between the two technologies.

Resolution Versus Speed

An object that prints with a layer height of 100 microns will take at least twice as long to print as the same object with 200-micron layers. That is because the printer has to print double the number of layers for the same object. The lower the microns, the longer to print, but the higher the print quality. Additionally, more layers mean more passes of the extruder, which means more heat gets put into the model. The model layers might need a slightly slower print time to cool off to prevent deformations.

The other metric for resolution is the XY accuracy—how accurate each individual layer (if viewed by itself from the top) can be "drawn" by the printer. Because most FDM printers use a .4 nozzle diameter (400 microns) the size of the "path" of deposited material coming out of the nozzle is around .48 (480 microns). This "path width" needs to precisely overlap adjacent paths to create the detail on a print. Therefore, at sharp points, some detail can be lost as you can see in **Figure 4-3**. While this still results in acceptable print quality, you will see in the resin printing chapter (Chapter 6), there are other more accurate technologies available if you need extreme detail.

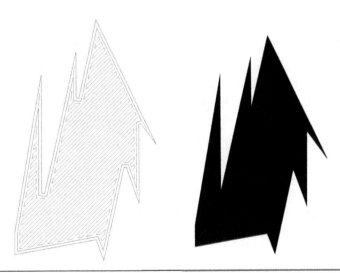

FIGURE 4-3: XY positional accuracy between layers created using FDM (left) and resin (right). Notice the sporadic blue lines of the extruder trying to fill in space on the FDM printer at the sharp areas, while the SLA print shines crisply, creating perfectly pointed shapes.

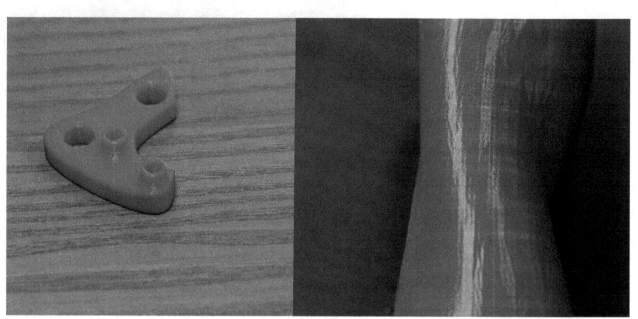

FIGURE 4-4: Bracket with no gaps, and perfect layer finish (left). Very smooth 3D model of "equation of time" based on work from LongNow.org (right).

FDM printers do some phenomenal work, though! Look at some examples in **Figure 4-4** to see the quality level achieved on FDM 3D printers.

FRAME/CHASSIS

This is what you first see when you look at a 3D printer; the overall general shape. There have been many 3D printers that have enjoyed crowdfunding success on websites such as Kickstarter or Indiegogo because of beautiful 3D printer enclosures/frames. If the enclosure looks nice, people are more inclined to infer it's a quality printer. This is only somewhat true, and this is where your own research can really help you make an informed purchasing choice.

The frame/chassis needs to be structurally sound over time. 3D printers have moving parts, they get bumped and prodded when you're removing prints from the build plate, and they need to deal with inherent vibrations from the printing process itself. There are few (if any) printers in the market right now that use a wooden frame, so best to stay clear of any you might see. The same thing goes for a plastic frame...you will always want to look "inside" of a printer to make sure that there is a rigid metal frame underlying any fancy exterior to help long term durability. Additionally, If you will be moving your printer from location to location, frame strength is something you will especially want to keep in mind, and a metal frame is more solid than wood.

Don't judge a printer by its cover! Many 3D printers have 3D printed parts holding them together. And why not? It helps keep costs down and allows for easier upgrades later on. While 3D printed parts might lack the durability of machined parts, having 3D printed parts in your printer should not be a mark against the overall printer itself. If an update to the design becomes available, you can print your own upgrades! 3D printed parts can be found in all price ranges of printers, and some have no 3D printed parts at all. As with all products, it is advisable to read the forums and unbiased reviews to see if a printer, no matter how it is assembled, is reliable.

There are quite a few "kit" 3D printers that you have to put together yourself that are incredibly structurally sound. One standout printer is named the Prusa i3 MK3S kit from prusa3d.com which positively answers most of the bullet-point questions below and stands out as one of the best and most popular 3D printer kits on the market (as long as you are willing to devote 6-10 hours to assemble the printer yourself, following detailed online guides!).

WHEN YOU ARE LOOKING AT FDM PRINTERS (OR REALLY ANY 3D PRINTER), LOOK CRITICALLY AT THE CHOICES THAT THE MANUFACTURERS MADE:

- How thick is the sheet metal forming the chassis? Does the chassis have any flex if you try and bend it?
- Is there flex to the horizontal arm that carries the extruder assembly which would result in potential alignment issues? (There should be no flex)
- Does the printer have a heated bed? (You definitely want a heated bed)
- What sort of covering does the printer use on top of the heated bed? (Many printers have removable build plates which make print removal very easy)
- What sort of components went into the extruder assembly? Are they machined parts, are they custom made, or something that is shared by many other manufacturers? (For example, extruders by the company "E3D" are excellent, and used in many 3D printers)

We have a Prusa i3 MK3S (with the additional 5 filament "multi-material upgrade v2") at the office, and it is a go-to machine we use for our client 3D prints. When the RepRap Project started developing, many of the 3D printer development lines were named after famous geneticists: Darwin, Mendel, etc. An individual maker named Josef Prusa was very active in that community and developed an entire line of printers that is named after him! There are clones of the "Prusa" printer all over the internet, but they are not as reliable as ones purchased directly from Josef's site.

These printers are hallmarked by easy print set up, an elegant two-piece chassis, and great build quality. The Prusa printer's thin metal frame might not look robust, but a lot of thought has gone into making this a reliable printer. There are many printers with the "Prusa i3" name out there, but we recommend the official Prusa i3 MK3S from the prusa3d.com website, not only because they are great printers, but imitation printers reduce their cost in notable areas such as power supply and extruder quality. This results in a less expensive printer, but at the cost of a more error-prone print process. Prusa is a major contributor to the

FIGURE 4-5: E3D hotend on Prusa i3 MK3 printer (arrows pointing to heat break with fins, and nozzle block, covered with blue silicone heat-retention "sock"

software that runs and "slices" prints, and it is good to support companies that donate back thought leadership that help the global 3D printing community.

Costing around $750 for a kit or $900 for a pre-assembled printer (all amounts in USD), these Prusa printers are right in the top positions of consumer-grade 3D printers. Because they are open source, you can download the design files yourself and make the printer better over time (if you think you can -- they're already pretty great).

As an example, Prusa printers do not feel they need to reinvent the wheel...or in this case, the extruder. The Prusa printers use the excellent hot ends from a company based out of the UK, named E3D (**Figure 4-5**).

These 3D printers also incorporate cutting edge technology, like the ability to print in up to five materials in a single print run, which at the time of this writing, no other consumer printer on the planet can do so seamlessly (**Figure 4-6**). This multi-material setup allows for different colors, different materials (PLA and ABS

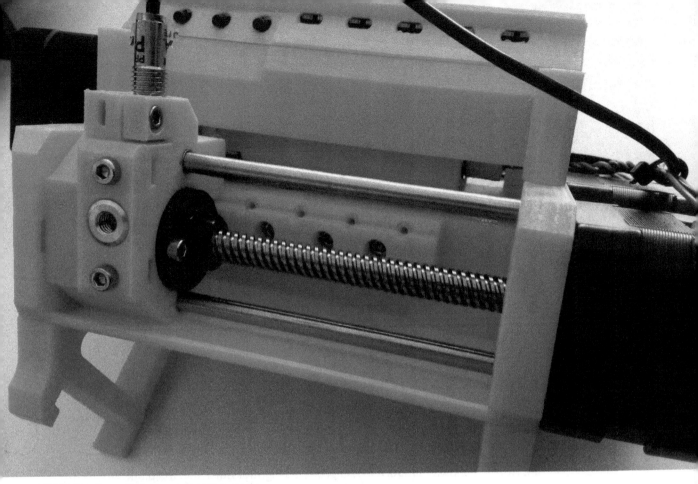

FIGURE 4-6: Prusa i3 Multi-material 2.0 upgrade

and soluble material all in one model), with the only constant factor being the filament diameter (1.75mm). A word of advice: If you are just getting into 3D printing, go with the single material printer first...you can always upgrade to the more complex, five material upgrade later on.

The two figures just shown illustrate that even if a printer's design is completely open to the world, a robust and quality product can still succeed in the marketplace.

Remember, over 90% of companies in the consumer 3D printing realm are what would be considered "small businesses" with 40 or fewer employees, so the fit and finish of each product is something that characterizes the company behind the product. Don't be fooled by fancy renderings or pictures; there is no substitute for seeing and experiencing the build quality yourself (or in reading independent reviews written by knowledgeable reviewers).

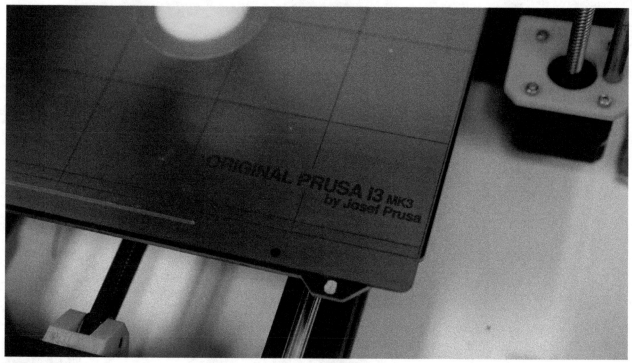

FIGURE 4-7: PEI film covering a metallic build plate to help with layer adhesion

BUILD PLATE

The Build Plate (or Print Bed), as shown in **Figure 4-7**, is pretty straightforward. It is the place where the 3D model is built, layer by layer. Different manufacturers use different materials for the build plate, and you will commonly see acrylic slabs, glass, garolite (a type of fiberglass board useful for nylon filaments), and metal build platforms.

In the image below, a PEI (Polyetherimide) film has been applied to a magnetic build plate that can be removed easily from the printer's heated bed. This film helps the first layers stick to the build plate. Making sure that your 3D print sticks to the build plate is one of the most important parts to the print process and one that you should evaluate carefully before you purchase an FDM printer.

All FDM printers require the build plate to be level to the extruder assembly's movement plane(s), otherwise, your prints will print lopsided or, worse, fail completely. Some printers, like the Prusa i3 MK3S, have an "auto leveling" probe that reads the build plate and adapts the print to compensate dynamically for any skew to make bed leveling extremely easy (and automated).

Many people think that the term "leveling the build plate" means that the plate needs to be level to the ground. This is not the case! The build plate needs to be level to the travel of the extruder assembly for the print itself to be level. In other words, the extruder's movement relative to the build platform needs to be parallel in order for prints to be successful. As long as you don't skew critical parts, the printer can even print upside down!

EASE OF CALIBRATION

When evaluating 3D printers, it is important to research how easy it is to calibrate or level the build plate to the extruder assembly. Today, many FDM printers use automated systems to level the extruder to the build plate. Whether you have an automated leveling system or not, you will be responsible for making sure the print bed is leveled at the start of every print.

It is also good to understand that build plates themselves are different. Some are heated, which allows for more materials to be printed on them, and some are non-heated. The heated build plate helps more heat-sensitive materials with adhesion of the print to the build plate by keeping the print warm as it prints. The extruded filament is extremely hot and without a heated build plate, the new layers are much warmer than the cooled-off previous layers, leading to contraction of the print, and adhesion issues with the print popping off of the plate, causing failed prints. A picture of a heated build plate is shown in **Figure 4-8**.

You can print a wide variety of materials without a heated bed. Of course, having the option to use a heated plate opens up your material selection to more materials, but usually comes at an added cost to the base printer price.

FIGURE 4-8: A heated print bed, on top of which a metal print surface is later attached.

Getting 3D prints to stay adhered on the build plate is an absolute requirement for a successful print. If the print starts to peel up off the plate while the print is in progress, your print will come out lopsided, or completely pop off the plate altogether! Some build plates (like in the Prusa i3 MK3 printer) come with a coating on them that encourages prints to adhere well to a heated build plate. If you do not have a coated build plate, you must prepare your build plate in order to get the 3D print to stick to it. Some people use blue painter's tape, while others use a simple school-crafting glue stick; there are many ways to get the print to adhere to the build plate. A more comprehensive description of these techniques is described in the Troubleshooting FDM Prints chapter section (Chapter 5).

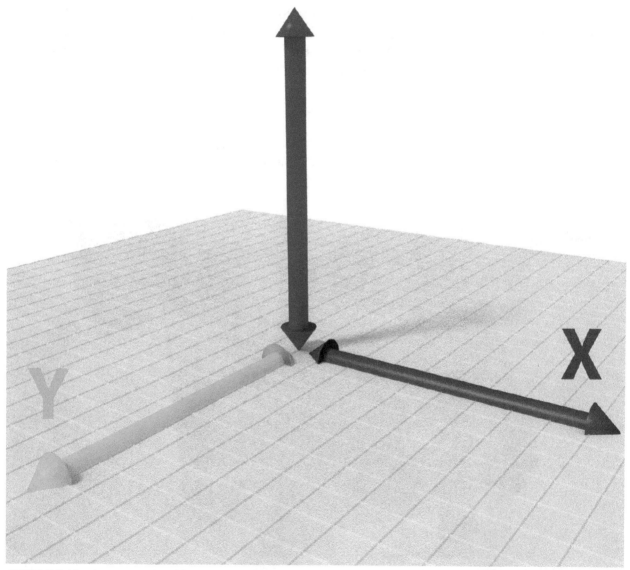

FIGURE 4-9: Visual representation of Cartesian coordinates with X, Y, and Z axes labeled

LINEAR MOVEMENT CONTROLS

All 3D printers need to use motion to print. There are two general types of 3D printing movement systems, both achieving good build quality but going about the process in very different ways. The two types are called Cartesian and Delta printers.

Cartesian movement systems (**Figure 4-9**) were the first to develop, and their movement is based on principles you probably learned in school.

THERE ARE THREE AXES IN A CARTESIAN PRINTER—X, Y, AND Z:

- X movement is to the left and right.

- Y movement is to the front and back.

- Z movement is up and down.

These axes can change between printers, though. Some printers will move the nozzle in what seems to you to be a left-right movement, but call that plane of movement the Y plane. Suffice to say that all Cartesian printers use the X/Y/Z coordinate system, drawing out each layer in the XY space, and then moving the extruder assembly by a certain number of microns in the Z direction to start the XY print process over again.

Delta printers go about 3D printing in a different way. These printers borrow from mechanisms that have long been used on assembly lines called "pick and place" machines. Instead of an X/Y/Z linear system, they use a system based on a "floating" extruder assembly that is pulled in a combination of directions by three arms that are attached to linear rails and pulleys, as shown in **Figure 4-10**.

FIGURE 4-10: The Rostock Max v2 Delta printer from SeeMeCNC. Picture was taken looking down from the top of the printer to the build plate. The extruder is in the center

To understand how a Delta printer works, imagine you and two friends holding a hockey puck with a pen in the center of the puck. The puck is in the center between all the people. If you all coordinate your movements, you can draw pictures with that hockey puck, but you all have to work together to get the hockey puck and pen in the right place at the right time. That is how Delta printers work. If you are having a difficult time visualizing the movement, do an Internet search for "Delta printers" or "pick and place machine" and watch some fascinating videos.

Why were Delta printers created if the Cartesian system was developed first and worked well? Delta printers overcome two major problems the Cartesian systems have. First, Cartesian 3D printers are limited in Z height (the overall vertical height of the print) to the maximum length of the large lead screw that raises or lowers the build plate, or extruder assembly as shown in **Figure 4-11**.

Lead screws are expensive, and as they grow in length they become more and more variable in how straight they are. Delta printers don't need to worry about the height of prints because the central extruder assembly glides up on smooth rails pulled by pulleys. Long and smooth linear rails are widely available, so Delta printers are almost always capable of far taller prints than their equivalently priced Cartesian printers.

FIGURE 4-11: Photo highlighting one of two lead screws in the Mendel Max 2.0 printer

Second, Delta-style printers often enjoy higher printing speeds than Cartesian printers, generally because they try to eliminate weight from the extruder assembly as much as possible. While Cartesian printers usually have all of the filament pushing and heating mechanisms directly above the extruder, Delta machines use what is called a Bowden-style extruder to drive the filament. In a Bowden system, the heated block and nozzle are indeed close to the build plate, but the stepper motor that grabs and pushes the filament is located somewhere else, off of the central area. This allows for the movements of Delta printers to be faster because they do not have to push all that weight around.

The drawbacks to Delta printers are twofold (these might not be drawbacks to you, so judge accordingly): Delta printers print on a circular build plate because of how the three extruder arms move. Accuracy at the edges of the circular build plate tends to drop because one or more of the arms is being pulled to its maximum length, which introduces abnormalities in the print dimensions. Many Delta printers avoid this issue by advising you to only print in a smaller central area of the build plate, or by enlarging the size of the entire Delta printer, and then making the outer areas "off-limits" to actual printing. So, while offering greatly enhanced Z height printing capabilities, Delta printers may have a smaller X/Y print area.

BOWDEN-STYLE EXTRUDERS (ON DELTA AS WELL AS CARTESIAN PRINTERS)

The second drawback concerns the use of the Bowden extruder itself. The filament is grabbed by a pinch-wheel and pushed through a guide tube (usually made of Teflon) from farther away, as shown in **Figure 4-12**. This is fine for most filament, but the amount of arc/curvature in the tube can cause problems when, for example, using flexible filaments. This is because the flex in those filaments gets increased over the long distance from the extruder to the hot end, and the filament can press through the heating block unevenly, causing irregular extrusions.

Just to make sure you understand that we are not biased against Bowden configurations: the printer pictured above, from Ultimaker (**http://www.ultimaker.com**) is one of the most successful 3D printers anywhere, catering to consumer and professional FDM consumers alike. It uses a Bowden-style extruder and produces phenomenal prints.

FIGURE 4-12: Bowden-style extruder mounted on the back of an Ultimaker 2 printer

EXTRUDER

The filament feeds through an extruder (or extruder assembly). The extruder is a collection of parts that feed the filament into a channel, heat up the filament to its melting point, and then force that molten aterial through a small-diameter nozzle onto the build plate.

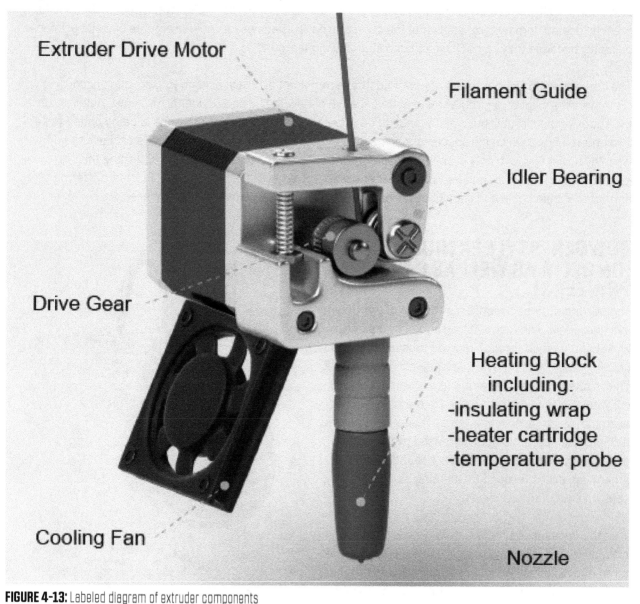

Extruder Drive Motor

Filament Guide

Idler Bearing

Drive Gear

Heating Block
including:
-insulating wrap
-heater cartridge
-temperature probe

Cooling Fan

Nozzle

FIGURE 4-13: Labeled diagram of extruder components

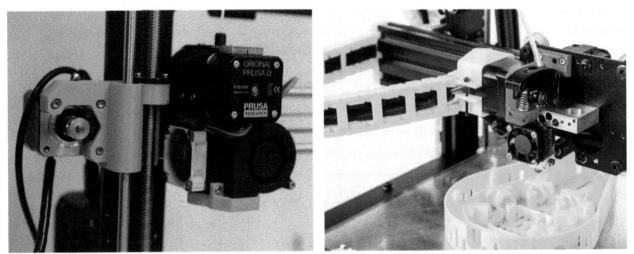

FIGURE 4-14 (LEFT): Prusa i3 MK3 extruder assembly **FIGURE 4-15 (RIGHT):** PMaxstruder from Maker's Tool Works on the Mendel Max 3.0 printer

THE EXTRUDER ASSEMBLY IS MADE UP OF SEVERAL PARTS, AS YOU CAN SEE IN FIGURE 4-13. THEY ARE DESCRIBED HERE:

- Drive gear (used for gripping the filament)
- Extruder drive motor (turns the drive gear)
- Filament channel (filament is guided through here)
- Idler bearing (pushes the filament against the drive gear)
- Heating block (melts the filament)
- Nozzle (through which the molten filament flows onto the build plate)
- Cooling fan (cools prints at the right time)

All FDM printers have variations on this setup. Some have extruder assemblies that are single units where all of the complexity is hidden inside one replaceable unit, while other 3D printers choose to go a more modular path to allow users to upgrade and change out pieces as their needs change or upgrades become available.

The images above demonstrate the different types of extruder assemblies from different manufacturers. **Figure 4-14** is an extruder from the Prusa i3 MK3S and **Figure 4-15** is of a self-assembled printer (called a Mendel Max 3) that combines machined and 3D printed parts. These two examples show slightly different approaches to performing the same task: pushing a strand of filament against a heating element in a controlled manner.

Generally, the only thing that you would consider changing is the size of the nozzle at the very bottom of the entire assembly. This is where the molten material exits the assembly to be deposited on the build plate. The width of the opening determines what sort of final characteristics your printed model will have.

Standard nozzle diameters vary but are generally around .4 mm. These dimensions dictate the width of the standard "line" of filament that comes out from the nozzle, called the extrusion width.

THE NOZZLE SIZE DICTATES A FEW IMPORTANT THINGS:

- A finer nozzle allows for thinner walls to be created and for surface features to be more defined (like using a smaller/sharper pencil for fine-detail drawings).
- Physics determines the maximum amount of material that can flow through a specific physical nozzle opening. If you want to print large objects quickly, then you will need a larger nozzle size.
- A larger nozzle size allows for higher layer heights to be created, with more material being extruded per "line." This results in decreased quality of surface finish, an increase in object strength, and a decrease in print time (sometimes half the time or less!

Figure 4-16 gives an example of the size of the individual extruded lines of material. A .4 mm nozzle usually creates a single extrusion line that is about .48 mm wide.

If you need to print objects that are more specialized (either larger or smaller), many printers offer a way of changing out the nozzle size to accomplish certain goals. A nozzle size of .25 mm is relatively slow at printing larger objects, but if you need to create very thin/small details on your model, this nozzle size would be your preference.

On the other hand, there are nozzle and extruder assemblies. such as the one from E3D called "The Volcano," which allows for much larger extrusion widths and heights, as you can see in **Figure 4-17**.

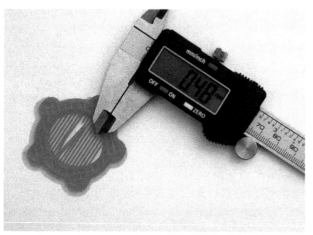

FIGURE 4-16: Calipers show an extrusion width of .48mm extruded from a .4 mm nozzle. Any nozzle from .35 mm to .45 mm is a standard size that is both fast enough to print large objects and small enough to print small features.

FIGURE 4-17: An object 3D printed by the E3D Volcano extruder, with .8 mm high layers. This object printed in half the time compared to 3D printers equipped with .4 mm nozzles.

FIGURE 4-18: The Olsson Ruby nozzle with an engineered ruby at the very tip.

The nozzle diameter that printed the object in **Figure 4-18** was 1 mm wide, and nozzles like this can print very large objects much more quickly than smaller standard nozzle sizes. If you want the flexibility of picking and choosing what types of nozzles you print with, you will want to consider choosing a 3D printer platform that will let you switch out nozzle sizes without too much hassle, or without voiding your 3D printer warranty.

Some boutique companies have started to cater to more specialized 3D print materials. A great example is The Olsson Ruby **(http://www.olssonruby.com)**. When you are printing with extremely abrasive materials like carbon fiber, or any of the metallic-infused filaments, the very act of dragging the filament through the nozzle itself can actually wear through standard brass nozzles! The Olsson Ruby nozzle is a bit pricey, but it has an engineered ruby as the actual nozzle, and is almost abrasion-proof, allowing you to print with any material without wearing down your nozzle. An image of the ruby nozzle is in **Figure 4-18**.

Many companies offer one warranty for the printer itself, and another, shorter warranty on the extruder assembly because of the faster wear and tear potential. Typically, both the nozzles and the drive gear (the toothed gear that grabs the filament and pushes it through) are considered "consumables" because they experience the most wear of all 3D printer components. The rate at which you need to replace those parts, however, could vary between once per year to once every several years, depending on how you use your printer, what filament material you use, and how well you maintain the printer. If you are going to be 3D printing a lot on an FDM printer, and experimenting with various types of materials, it is a good idea to see how easy it is to swap out or clean those replaceable components.

FILAMENT

One of the core components of an FDM printer is the thin strand of filament material used to make an object. As we compared in a previous analogy, filament used for 3D printing can be seen as similar to the "ink" used in a 2D printer. Filament comes in two standard sizes, which describe the diameter of the filament: 1.75mm and 2.85mm (sometimes sold as 3mm in diameter). Your 3D printer will support only one diameter of filament, and you cannot easily use a different size filament if your printer was not designed to use it. Fortunately, there is usually no appreciable price difference between filament diameters, though 1.75mm is a more common filament diameter size. From a technical standpoint, 1.75mm is better (and more common) because it requires about one-sixth as much force to melt and extrude than the thicker strand.

Thinner strands require less bulky, expensive motors to drive the filament through the extruder, which lessens the overall mass of the moving print head. Alternatively, if you are going for a very large nozzle size (1mm and above), then a 3mm diameter filament printer will be able to extrude more filament in a quicker fashion and will be a vital requirement for larger-scale printing.

There is an ever-expanding range of different materials that can be fed through consumer 3D printers, with the FDM printers offering the most selection out of all the printing technologies discussed in this book. The most common material is called polylactic acid (PLA) and is actually a type of non-edible carbohydrate (sugar) derived predominantly from corn. This material prints beautifully and is the starter material of choice for most consumer 3D printers for many reasons, including less smell produced while printing. Both economical and easy to find, it's a good choice for beginners and school settings.

The second most popular filament is PETG (polyethylene terephthalate, glycol-modified). While the petroleum-based ABS (acrylonitrile butadiene styrene) used to be the favorite "more durable" filament, PETG has surpassed ABS in popular use. PETG is chemically stable, does not off-gas noxious fumes (as ABS does), and is more resistant to outdoor factors, such as UV (ultraviolet) radiation.

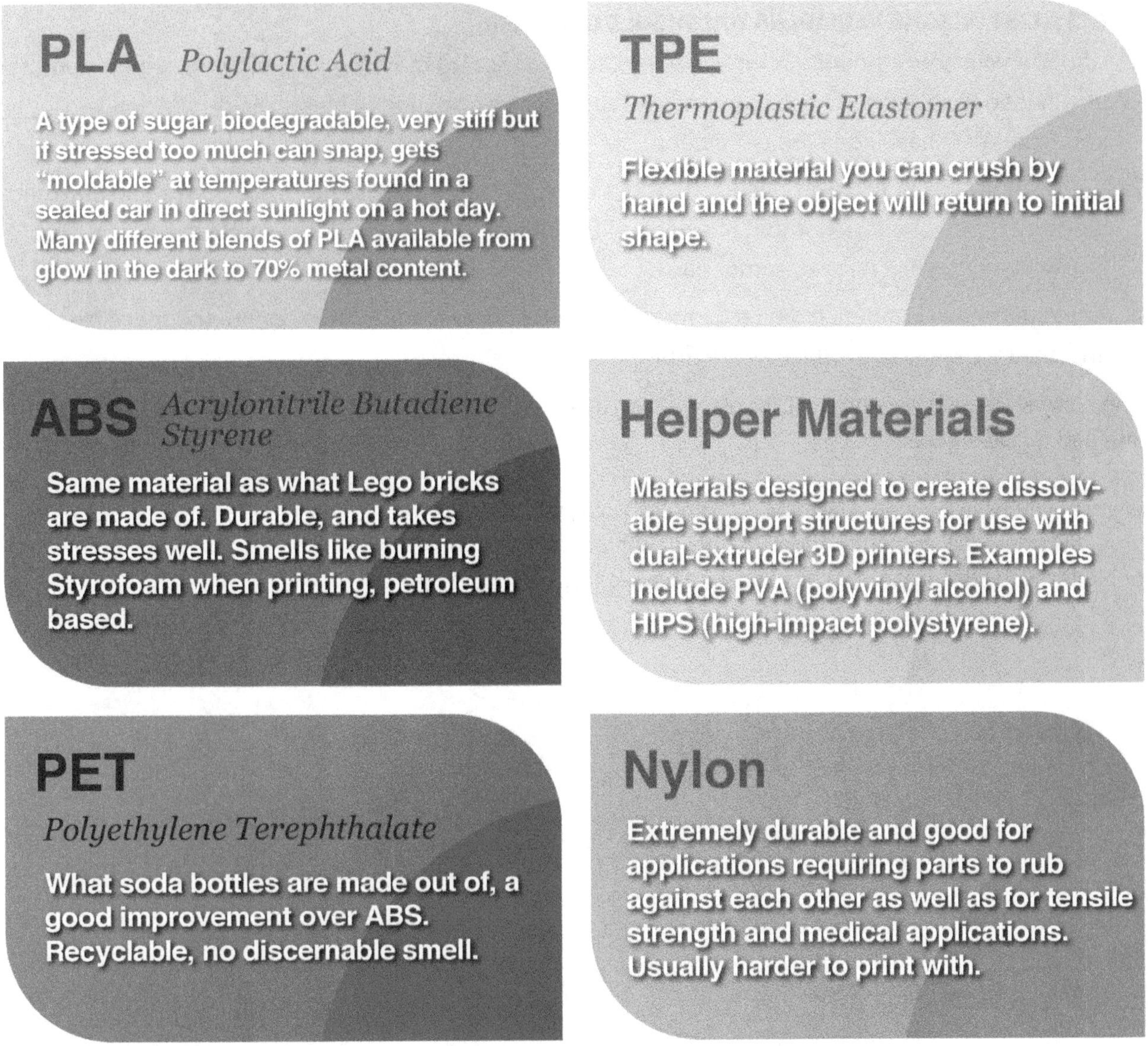

PLA *Polylactic Acid*

A type of sugar, biodegradable, very stiff but if stressed too much can snap, gets "moldable" at temperatures found in a sealed car in direct sunlight on a hot day. Many different blends of PLA available from glow in the dark to 70% metal content.

TPE
Thermoplastic Elastomer

Flexible material you can crush by hand and the object will return to initial shape.

ABS *Acrylonitrile Butadiene Styrene*

Same material as what Lego bricks are made of. Durable, and takes stresses well. Smells like burning Styrofoam when printing, petroleum based.

Helper Materials

Materials designed to create dissolvable support structures for use with dual-extruder 3D printers. Examples include PVA (polyvinyl alcohol) and HIPS (high-impact polystyrene).

PET
Polyethylene Terephthalate

What soda bottles are made out of, a good improvement over ABS. Recyclable, no discernable smell.

Nylon

Extremely durable and good for applications requiring parts to rub against each other as well as for tensile strength and medical applications. Usually harder to print with.

FIGURE 4-19: Chart of the most common FDM printer materials (infographic by HoneyPoint3D™)

Several common filament materials are described in **Figure 4-19**, but note that some of these materials may or may not work with your 3D printer. It is best to check with your 3D printer manufacturer to see which materials they support.

There is a wide variety of filament available, ranging in price from $12/kg to $70/kg. This wide range in price does not always yield a better filament, so we advise you to rely on research and product reviews to inform your choice in materials.

HERE IS A LIST OF SOME VARIABLES YOU SHOULD CONSIDER:

- Check if your printer manufacturer has tested the filament that they sell with the printer you purchased from them. These filaments will usually be at a slightly higher price, but you will know that they have done the testing for you.

- Find online sites with user reviews. If others have printed well with filament from a specific manufacturer, then chances are that filament will print well for you too. Most reviews are real, but like with any review format, some may be fake.

- Try to purchase filament from large manufacturers. They usually have conducted more testing and are less likely to produce faulty filament.

Long story short: Having a great 3D printer with an awesome 3D model will not be able to save a failed print from bad quality filament.

Now that you have learned how FDM 3D printers work, let's discuss the reality of the 3D printing learning curve: 3D prints fail, and quite often. The next chapter reviews some common print problems you might encounter with consumer FDM printing, and how to troubleshoot them.

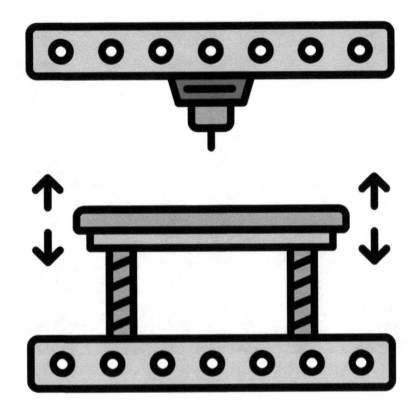

CHAPTER

5

TIPS FOR SUCCESS IN FDM PRINTING AND VISUAL TROUBLE-SHOOTING GUIDE

FIGURE 5-1: Spools of filament stored in a mostly airtight container, with desiccant silver rectangular pouches added

This chapter is meant to reiterate one point: 3D printing takes time to master and when you are practicing your skills, 3D prints will fail, sometimes quite often. Even though the authors of this book are considered experts, we still get failed prints from time to time: sometimes the filament or resin went bad, sometimes the first later didn't stick to the build plate. When we first started our 3D printing journey years ago, we had a lot more 3D print failures, and this chapter is intended to help you learn from our mistakes so that you can hopefully shorten your learning curve.

In this chapter, learn about specific details on how to set up your 3D printing-related tasks before you even start printing. Later in the chapter, you will be able to visually see our guide to common print problems. This is a newly-added chapter from the first edition of this book, as readers asked for some good starter advice. Let's start with what you need to know before you begin the printing process.

BEFORE YOU START PRINTING

All FDM printers use various types of filament. We discussed what filament is in the previous chapter, but we need to state one point here: **Proper storage of your filament is key.** Success in 3D printing is about **reducing the number of variables that change between prints so that your settings are successful each time**. In terms of filament, between prints you want your filament to remain as new as when it was first delivered to you.

Make sure that your filament does not accumulate dust and debris, because they will then get drawn into

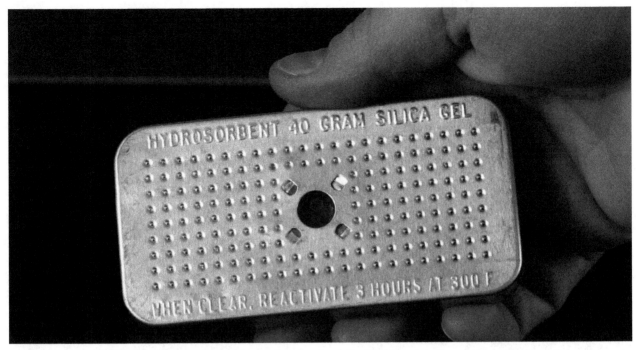

FIGURE 5-2: Desiccant canister displaying instructions on how to refresh the humidity-absorbing material inside the metal case

your nozzle, clogging it. Storing filament in a clean, temperature-controlled environment is one of the best ways to protect your filament when not in use.

Many FDM materials, like PLA and nylon, are **hygroscopic** materials, which means they absorb moisture from the ambient air. If you live in a humid area, this could be even more of an issue for you than dust. For all the reasons mentioned, it is best to store filament you are not using in an airtight container like the one shown in **Figure 5-1**. The desiccant canister you see in **Figure 5-2** is inexpensive, can be purchased online, and can be "recharged" if it absorbs too much water by drying it in a kitchen oven for a few hours. Two to three of these desiccant pouches per container are ideal for filament preservation.

BUILD PLATE ADHESION

In the 3D printing world, there are many instances where users have discovered some new and innovative ways to make 3D prints more successful. Thankfully, these finders of knowledge then spread that information to others in the community through forums, user groups and media. Experiencing and finding different ways to get 3D prints to stick to the print bed is one such area many people have focused on. A good tip is to first clean your fingers that can get on the plate with normal isopropyl alcohol to get the oil off.

The following are some techniques that people have discovered to successfully enhance build plate adhesion:

FIGURE 5-3: PEI film applied over a glass build plate

PEI COATING OR SHEET

PEI is a specific type of material which is sticky to most 3D print materials (For example, PLA, ABS, and PETG). Keep in mind, with a heated bed, you generally would not need to use any type of adhesion technique -- you can just start printing.

Note though: PEI sticks TOO well to some materials (like flexible filaments), so you would need to use one of the below methods in addition to PEI for those materials. The slight amber color in **Figure 5-3** is the PEI film with a sticky back that has been applied over a build plate. Once applied, the PEI material should last for over a year of constant printing, if not longer.

BLUE PAINTER'S TAPE

Applying this (blue) tape to the top of the print bed, as shown in **Figure 5-4**, has been shown to be a good way to get prints to stick. This tape is still removable, and sometimes the tape can peel up while printing. Unfortunately, it can take your model with it, so make sure the edges of the tape are pressed down well to prevent peeling. If you are lucky, it will not rip off when you remove your completed print from the print bed, allowing you to re-use that tape for the next few prints.

FIGURE 5-4: Blue painter's tape applied to the print bed in even, parallel lines

FIGURE 5-5: Consumer glue stick used to adhere prints to the build plate

WHITE GLUE STICK

As shown in **Figure 5-5**, this is a great method to use: it is much less expensive than even blue painter's tape. The downside is that it requires you to wash your print bed with a wet paper towel every few prints to remove excess glue buildup. And if your build plate is non-removable, then you need to be more careful, because you run the risk of dripping water on sensitive electronic components that are underneath. This type of glue is water-soluble, so if you live in a humid climate, or where it rains a lot, it will not perform as well.

FIGURE 5-6: Kapton tape that can be purchased in various widths

KAPTON TAPE

This is a transparent orange, special type of high-temperature tape that was originally designed for use in NASA space missions. It can handle the high temperatures of heated print beds, and is tacky enough for some materials to stick directly onto it. Kapton tape is primarily used in conjunction with a heated build plate to allow materials like ABS (and to a lesser extent PLA) to increase print adhesion to the build plate. A thin roll of Kapton tape is shown

FIGURE 5-7: Specialized Buildtak adhesive surface applied to the build plate

in **Figure 5-6**, with many different sizes available to purchase. Typically you would want to purchase as wide a strip as you can get for ease of installation, and to minimize the number of tape under or over lapping "edges" on your build plate (which will show up on the bottom of your prints).

SPECIALIZED 3D PRINTING BUILD SURFACES

These are sold as third-party add-ons that help adhere prints to the build plate, as shown in **Figure 5-7**. Typically, these all covering surfaces claim that no other adhesive or maintenance is needed other than infrequent cleanings. We have found that they work, but care should be taken to make sure you have proper nozzle alignment to the print bed. Otherwise, you run the risk of the heated nozzle melting one or more grooves into the print surface, necessitating a replacement surface.

HOMING AND LEVELING THE BUILD PLATE

One of the most important factors in encouraging your prints to print successfully is making sure that the first layer goes down evenly and "sticks" to the build plate properly. Many printers have automatic build plate leveling calibrated through an automatic probe on the printer, but if your printer does not have this function, you will need to manually level the bed using screwdrivers, screws, and patience. We highly recommend you work with 3D printers that have this automatic bed leveling feature.

In automatic bed leveling, there is a sensor next to the nozzle that actually does not set the height of the print bed, but rather reads the angle of the bed and adjusts the print process to compensate for a possibly skewed build plate. Remember, your 3D printer prints at layer heights as thin as one-tenth of a millimeter

FIGURE 5-8: Nozzle and sensor, with sensor highlighted in red

(100 microns). This is a very small distance! Even small irregularities in the tightness of screws or tiny variances in manufacturing tolerances for your printer components will cause a problem with bed leveling. Having the automated probe "read" where the corners of the build plate are, in relation to the nozzle, prevents one (or more) sides of your print from being shorter than the others. See a picture of a nozzle and sensor in **Figure 5-8**.

Even for printers with this feature, you still have to set the center height of the build plate manually by using a piece of paper as a guide to the proper "home point" for the nozzle above the print bed. The bed itself could be a little off at the edges, but due to the sensor, the print will still come out fine. See below for images on this process and look up videos on the Internet if you want a better visual of this process.

If you do not have this automatic bed leveling probe, your build plate will need to be adjusted before every single print. Even removing a print from the build plate can cause slight changes in the alignment of the build plate, thus you will need to do this process often.

THIS INVOLVES TWO STEPS:

- Leveling the bed so that all corners create a flat surface in respect to the movement of the left-right (X) and front-back (Y) axes.
- Adjusting the height of the bed so that the first layer lays down perfectly... not too far away from or too close to the build plate.

FIGURE 5-9: A view from underneath a manual-leveling printer, showing the four bed-leveling screws with 3D printed handles.

FIGURE 5-10: Images showing a piece of paper being slid between the nozzle and build plate with the nozzle being too far away (left) and too close (right)

Both of the following examples are from a manually-adjusted 3D printer. **Figure 5-9** shows the four screws on each corner underneath the build plate that need to be adjusted to make the build plate level with respect to the movement of the nozzle on the XY axes. Manual leveling printers are slowly disappearing, because auto leveling printers are more user friendly and in demand. The manual leveling images below are shown as an example of this older type of 3D printer feature.

You will test if the build plate is level by manually moving the extruder back and forth over the plate and see if the distance between the nozzle and the plate changes at different areas. Once the build plate is level, it (typically) should be about 100 microns, or .1 (one tenth) of a millimeter away from the nozzle, where the nozzle reads it as its "zero" point.

A great way to test this is to "home" your nozzle to where it thinks "zero" is, and then try to slip a sheet of paper between the nozzle and print bed. If you can barely slide a normal piece of paper underneath the nozzle, then it's homed. If the nozzle is too close to the plate, the paper will drag significantly under the

FIGURE 5-11: One foot of a 3D printer not fully seated because it's resting on stepper motor cables (rear oval). The front foot is floating in the air because of the problem at the back.

nozzle or bunch and not pass underneath at all. If the nozzle is too far away, the paper will glide underneath with minimal blockage and you will need to make adjustments. See Figure **Figure 5-10** for too far and too close "distance" of the up-down (Z) axis.

It is also important to note that even though your printer might have a level build plate, if the table or desk the printer is sitting on is wobbly or uneven, that can affect print quality over time as well. A well calibrated FDM printer, could fabricate objects when it is turned sideways, or even upside down! In fact, there have been people who have placed 3D printers on their backs and printed objects while walking around a Maker Faire!

Follow Manufacturer Instructions

Our instructions apply to many printers, but yours might be an exception. Always follow your specific printer manufacturer's instructions on how to level the bed and to set the Z height at which you should be printing.

In the real-world however, the chassis of a 3D printer is prone to warping over time if the printer is not solidly and flatly placed on a table or desk. See **Figure 5-11** showing a less desirable situation where one foot of a 3D printer is not contacting the table fully.

SLICER PROGRAMS

A 3D model, created using some type of CAD/CAM software, is usually represented in the computer by a collection of triangles arranged in three dimensions but your 3D printer needs to think of that object as a collection of flat layers, each a fraction of a millimeter thick. Before you 3D print any object, you will first need to "slice" it.

The slicer is a type of software that translates the 3D model's geometries into a precise pattern of motion for the extruder to build your model. The slicer also provides various settings like print temperature, speed, and, if necessary, generates support structures that are designed to keep overhanging parts of your model from drooping. You'll find a more comprehensive discussion on support structures in Chapter 11.

Most 3D printer manufacturers will either have their own slicer, or suggest a slicer to use, and we recommend that you start with the one recommended by the manufacturer. You might get better support from the user groups and from their consumer service if you use the one they recommend. In the instructions, you will find profiles provided for your printer and for the filament that comes from the manufacturer.

NOT ALL SLICERS ARE CREATED EQUAL

As with anything, some slicers are better than others. You may "slice" one model with specific settings in one slicer and then use another slicer with the same settings and get very different end-print qualities. Some slicers work for both FDM as well as resin printers, while some are specific to one type of printing technology.

THE FOLLOWING IS A LIST OF QUALITY SLICERS FOR FDM PRINTERS:

- Cura: free
- Slic3r: free
- PrusaSlicer: free
- Repetier Host: free
- KISSlicer: free and paid
- Simplify3D: paid only

Unless your printer manufacturer has a proprietary slicer, you have a range of slicers to choose from. Paid slicers sometimes have more developed features, though free slicers perform the slicing job perfectly fine as well. **Figure 5-12** shows tool path visualizations (layer patterns) from Simplify3D and Cura, comparing how each would print the same object.

Any of the free slicers will work well as long as the 3D model you are slicing is a "good" one without errors. While all of them can do some 3D model healing and fixing, you really want to make sure your model is good from the start. In Chapter 11, we'll discuss how Autodesk's Meshmixer can be used as a tool to help evaluate 3D models for potential errors.

To summarize our thoughts about slicers: it's best to stay with the slicing method your printer manufacturer recommends, but for a great many 3D printers, you are also able to use other slicers if you desire to have different/more control of the settings and layer patterns.

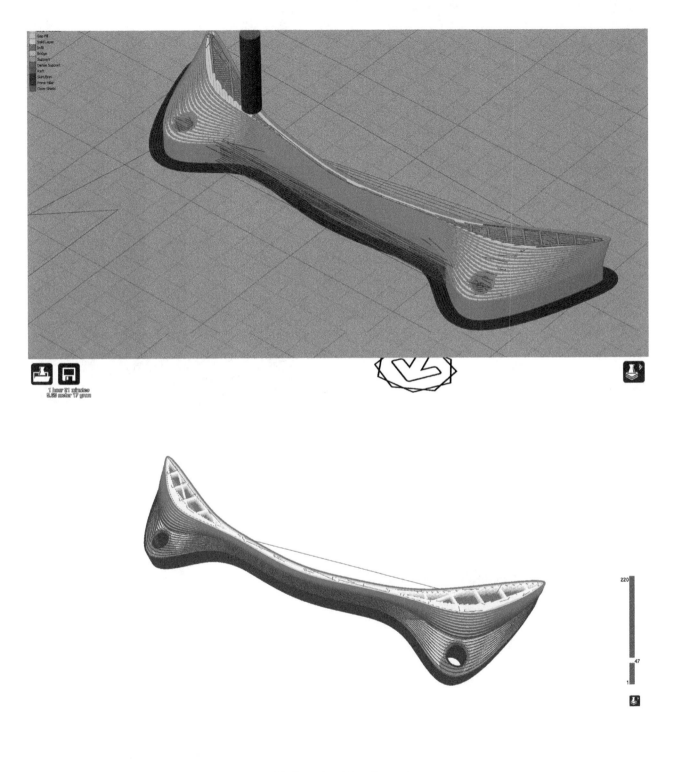

FIGURE 5-12: Comparison of tool paths between Simplify3D (top) and Cura (bottom)

FDM VISUAL TROUBLESHOOTING

This section provides a list of common problems that can occur with 3D prints, and suggestions on how to fix them (assuming the 3D model is good). Please remember that consumer 3D printers are still relatively "new" devices, and when a print fails, the printer will probably not even know it and the extruder keeps pumping out the filament. It has happened to quite a few people in the 3D printing community that left a 3D print unattended overnight to come back to a huge melted ball of filament on their build platform because the print peeled off the build plate and melted onto the hotend!

With good practices, and making sure to check on your prints fairly regularly, you will have great results! Now, let's look at some examples of what can go wrong, and how to fix those issues. Most all of the examples below will cause a 3D print to fail. There might be some examples that could be "ok" to let go and not cause you to cancel a print, and those will be called out specifically.

UNDER EXTRUSION / OVER EXTRUSION (CONSISTENTLY PRESENT THROUGHOUT THE PRINT)

FIGURE 5-13: Extrusion issues shown by a wavy surface pattern.

Symptom: The extruded lines of filament are either too thin, or too thick. **(Figure 5-13)**

Reason:

- Your filament might be out of tolerance (larger or smaller than 1.75 or 3mm).

- Your filament might need a hotter or colder temperature in order to flow more consistently.

- You might have some settings in your slicer not properly assigned.

Solution:

- ✅ Tolerance: Change your slicer's settings to the actual diameter of your filament (changing from 1.75mm to 1.65mm for example, if that is what you measured).

- ✅ Extrusion Percentage: Instead of changing your filament diameter, you can also change your extrusion percentage away from 100% down to 95% or up to 105% (as an example). With a lesser extrusion percentage, the printer will extrude slightly less material, thus lessening your overall flow of material.

- ✅ Temperature: 3D printers might have some variations in temperature. What one 3D printer thinks is 200ºC might actually be 215ºC on another 3D printer. Change your temperature down by 5ºC if your filament is oozing out of your nozzle (over-extrusion), or make the temperature hotter by 5ºC if your filament is not coming out consistently enough.

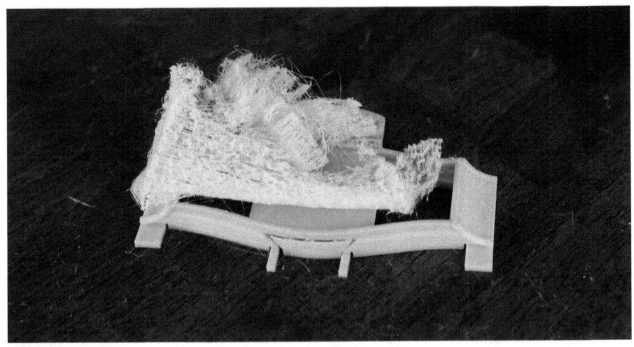

FIGURE 5-14: Transient extrusion issues over some areas of the failed print

UNDER EXTRUSION / OVER EXTRUSION (TRANSIENT / SOMETIMES HAPPENS)

Symptom: Certain layers on your print have noticeable gaps or completely fail, but other layers are OK, (Figure 5-14)

Reason: Not enough filament is coming out in those areas, but enough is coming out later on.

Solution:

- Barring a hardware issue with your printer like a loose connection, the filament is "out of tolerance" in just those areas, being smaller in diameter, and thus starving the print of material in those places. Consider buying filament from a manufacturer who is known for consistent filament diameter and quality.

- These problems can also happen if your filament has absorbed too much water during storage of the filament spool. Evaporation of that water during the print process can also cause inconsistent extrusion as the outer spool layers might have absorbed more moisture than the inner ones.

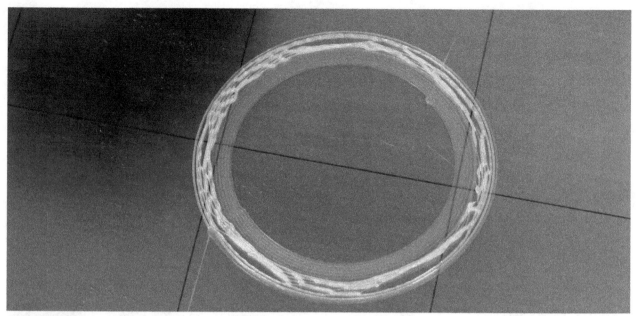

FIGURE 5-15: The First Layer is peeling off of the build plate. The inner rings are properly aligned, the outer rings did not adhere due to Z-height being too far from bed.

FIRST LAYER ADHESION ISSUES

Symptom: Your print is peeling off of the build plate **(Figure 5-15)**

Reason: Somehow, the adhesion between your print and the build plate failed, or failed over time.

Solution: There are many reasons why this can happen. Here is a list of things to check:

- Make sure your build plate has been cleaned with isopropyl alcohol or acetone before each print run.

- Make sure that your bed is levelled, and the Z-height is properly calibrated.

- Make sure your adhesion method is reapplied (put new glue stick material or painter's tape on).

- If you have a heated bed, make sure that is at an appropriate temperature.

- Some slicers allow you to print the first layer at a hotter temperature (10-15 degrees hotter) and at a slower speed. Consider enabling those settings.

- Check to make sure that there are no cold air currents blowing down onto your printer from an open window or an over zealous air conditioner. Environmental issues like that can cause one side of prints to fail, and might be hard to track down. Some people place wind barriers around their printer to help prevent this.

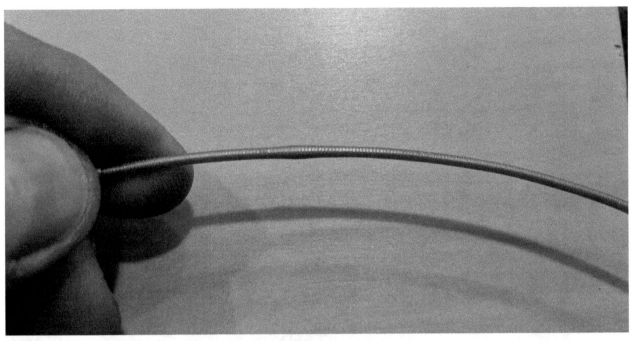

FIGURE 5-16: A sample of filament with an inconsistent diameter

FILAMENT STOPS EXTRUDING COMPLETELY

Symptom: Your filament has stopped feeding completely. **(Figure 5-16)**

Reason: There can be many causes.

• The pinchwheel might have spun in one area of your filament "biting" a channel into it.

• The picture above shows low-quality filament with a bulge in one area. This filament will not feed through the tightly-toleranced channels in your extruder.

• Having too low a temperature can also cause a jam, as the filament is not melting enough to feed through cleanly.

Solution:

✓ Ensure you are using high-quality filament from a manufacturer that has good reviews online.

✓ Make sure you are using the proper temperature for your filament. The "proper" temperature can be machine-specific, so some trial and error for a new filament is sometimes necessary.

✓ Take into consideration that if you are printing at faster speeds, you will need to raise your extruder temperature to handle the higher flow (usually by 5 degrees Celsius or so).

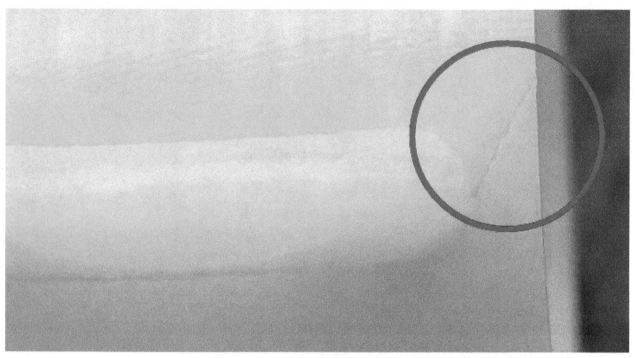

FIGURE 5-17: A slicing error caused by an error in the 3D model (highlighted in red).

NON-MANIFOLD 3D MODELS MAKE SLICERS GUESS (SOMETIMES INCORRECTLY)

Symptom: Your resulting 3D model does not look as it does on the screen. **(Figure 5-17)**

Reason:

- There were errors in your 3D model that the slicer tried to repair, but did not repair well.

- "Non-manifold" models refer to 3D models which are not fully "enclosed or sealed together" in their 3D form. Manifold models can be thought of like a perfect sphere made from a black plastic. Non-manifold models would be that same sphere but with a small hole cut out of the "skin" of the sphere, allowing one to look inside and see the interior geometry.

Solution:

- Unless you are supremely confident in your modeling skills, it is good to run your model through the Analysis → Inspector tool in Meshmixer to clean up any disconnected pieces and fill in any holes in your model. If you really want to be sure, follow up that operation with the Edit → Make Solid tool in Meshmixer. Please see Chapter 11 for a tutorial on Meshmixer.

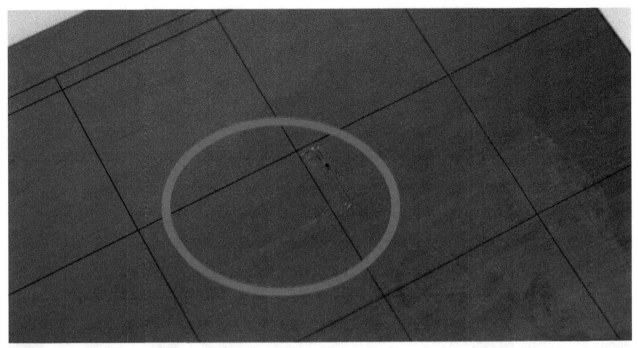

FIGURE 5-18: The aftermath of an improper Z-height highlighted in the red circle. The nozzle burned a channel into the PEI build material.

FIRST LAYER NOT APPEARING

Symptom: The first layer is not extruding any material, and the extruder is clicking **(Figure 5-18)**

Reason: The Z-height is set improperly.

Solution:

- This happens when the height of the nozzle is not properly set in the 3D printer, and the nozzle is completely blocked by the build plate. Run the calibration method for your printer to make the height of the first layer enough to allow material to flow.

FIGURE 5-19: The print is wobbly in some places. Note the jaws and teeth of this poor T-Rex print. Those areas were printed too hot, and suffer from heat build-up.

INCONSISTENT LAYER SURFACES

Symptom: Some areas of the print are uneven and thinly printed. **(Figure 5-19)**

Reason: That area of the print is getting too hot during printing.

Solution:

✅ Remember, the filament is heated to around twice the boiling point of water. If heat cannot escape from one layer of your print before another layer is deposited, over time, that heat will build up and cause your model to partially melt. Make sure that you enable your cooling fan on your prints (unless your specific filament's manufacturer recommends running without a fan), and some slicers also have a setting that will purposefully slow down a print if a layer print time is below a certain amount of seconds. In rare cases for very small objects, even that is not enough, and you might have to print one or more copies at a time just to give the print head some place to go while your print cools.

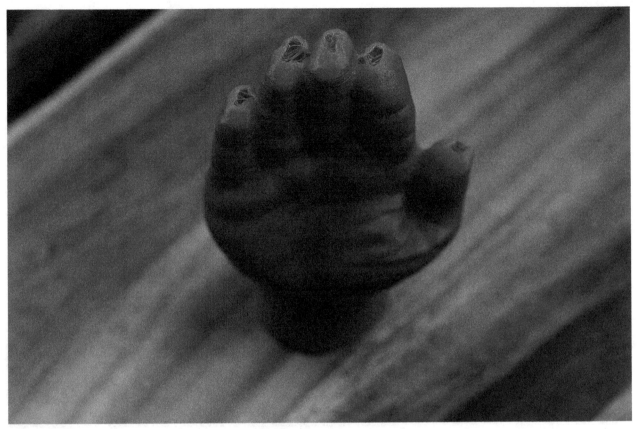

FIGURE 5-20: Example of a model of a hand that was printed hollow inside. The tips of the fingers needed support material underneath but did not receive it, thus there was drooping and an inconsistent surface finish.

DROOPING OR MISSING AREAS

Symptom: Parts of the object have drooping loops, or do not print at all. **(Figure 5-20)**

Reason: Support structures are needed under those layers because gravity makes things fall!

Solution:

☑ If parts of your model are drooping (as in Figure 5-20 above), then those parts need support material enabled in your slicer. The support structures are there to create a base from which those areas of your model with severe overhangs can be supported. Make sure to enable supports in your slicer. Some support material is removable and some support material remains hidden inside the model.

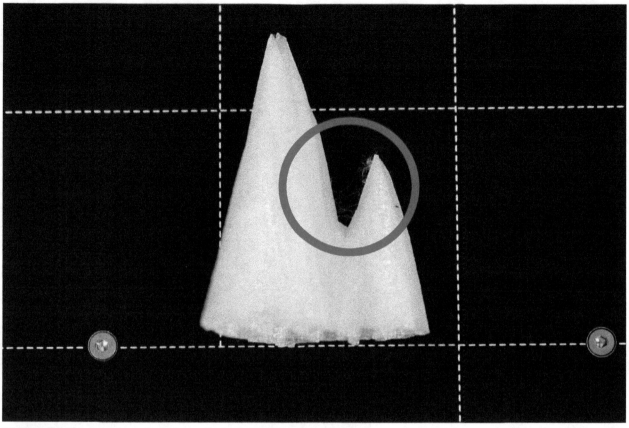

FIGURE 5-21: Thin hairs appearing between the peaks of this small mountain

HAIRY PRINTS?

Symptom: There are thin "hairs" sticking out from the side of your prints causing stringing. **(Figure 5-21)**

Reason: Too much filament is coming out of the nozzle when the nozzle moves to a new place, leaving a small trail of material behind.

✅ **Solution:** There are usually two interconnected reasons why this happens. If you are printing your filament at too high a temperature, the filament will ooze out of the nozzle too quickly, leaving a trail. The solution to that would be to decrease your temperature. There is also a setting in most slicer programs that is called "retraction." This reverses the filament back into the nozzle a slight distance before moving the nozzle to a new place, thus helping to prevent stringing. Either enabling retraction or choosing a slightly longer retraction distance can help eliminate this hair-effect. (Unless you like that look:)

Hopefully these common errors discussed in this chapter will be the majority of what you might encounter in trying to troubleshoot your failed prints. If you need more information, we encourage you to watch our online educational videos, reach out to people in the user forums, seek advice from the online printing groups and watch videos on YouTube about your specific issue. Keep reading to learn more about resin printers and how they work!

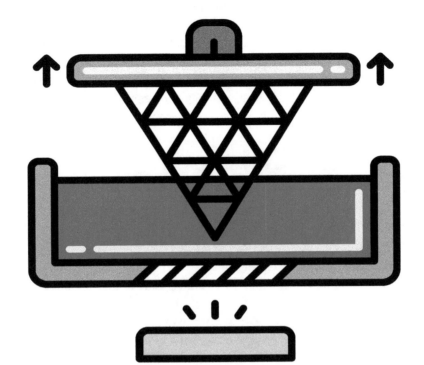

6

UNDERSTANDING RESIN PRINTERS

The previous chapter discussed FDM 3D printers, which represent about 90% of the consumer market. The other 10% belongs to a special class of 3D printers called resin printers. You might also see the term "SLA", masked SLA (mSLA), or DLP (digital light processing). SLA is an acronym for "stereolithographic apparatus," which basically means "gizmo that writes with light." These 3D printers create very detailed prints with smooth surfaces, but are more difficult to use, so it is important to know their benefits and drawbacks before you buy your first liter of resin.

HOW IT WORKS

These 3D printers do not use filament; instead, they use a liquid resin (polymer) that hardens or "cures" when exposed to ultraviolet light. The 3D prints from resin printers are still made layer by layer, but are created in a slightly different way.

The basic printing process for a resin printer is:
1. A vat of liquid resin sits in the 3D printer. The bottom side of that vat is transparent, allowing light to shine through.
2. A build plate gets submerged downwards from the top to the bottom of the vat, until there is a very small layer of resin between the bottom of the vat and the build plate.
3. A controlled light, pointing upward from the bottom of the 3D printer, hits the build plate in a specific pattern, hardening the resin in that specific pattern.
4. The build plate then moves slightly upward, pulling the newly cured layer along with it.
5. The light shines again in a slightly different pattern from the previous layer, and then cures to the layer that was created before it.
6. The process repeats until the object is complete.

For a graphical representation of the process, see **Figure 6-1**.

The goal of this chapter is to show you the differences between FDM and SLA printers, as well as point out some unique considerations when running an SLA printer.

HERE ARE SOME NOTABLE DIFFERENCES BETWEEN RUNNING AN SLA PRINTER VERSUS AN FDM PRINTER:
- The resin is ideally kept at around 80 degrees Fahrenheit in order for the viscosity of the resin not be affected. A thicker/colder resin will need more time to flow back under the build plate between layers.
- The resin has a noticeable odor...a kind of sweet, but strong chemical smell that may bother some people.
- The resin printer must be kept away from windows because any extra sunlight might cure some of the resin inadvertently.
- You must wear gloves when handling the resin and the newly created 3D print, as well as, for

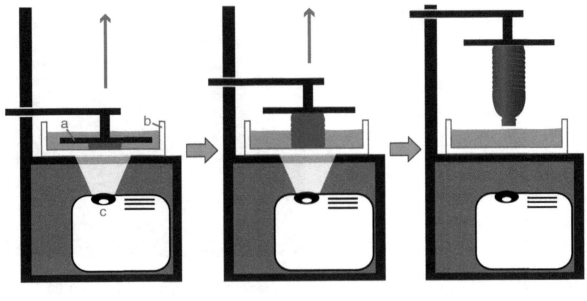

The build plate (a) starts out submerged in a liquid resin vat (b). A laser or a projector (c) shines light through the bottom of the resin vat, which cures the resin in a very precise shape, making it solid.

After each layer of resin is cured, the build plate is slowly raised upwards, peeling each cured resin layer off the bottom of the resin vat. The build plate is then lowered back down into the vat where the next layer will be cured.

Over time, the 3D print will take shape as it is pulled out of the liquid resin vat. Once the print is removed, excess resin must be rinsed off with isopropyl alcohol, followed by exposure to UV light in order to finish the curing process.

FIGURE 6-1: The resin printing process (shown using a DLP light source). Infographic by HoneyPoint3D™

cleaning the printer to avoid getting resin on your hands. This means, essentially, wearing gloves every time you physically interact with the printer.

- Kids and pets should be kept away from the 3D printer and resin. The resin should be considered as dangerous as household bleach.

- After the resin print is finished, it will need to be rinsed by hand in isopropyl alcohol.

- Once the resin print is washed off, it is a good idea to place the print in direct sunlight or in a UV curing box for 10 minutes to finish curing the outer layers.

- After the printing is finished, you will need to remove tiny cured particles of resin that may have appeared in the vat and could interfere with future prints. You should pour the leftover resin through a commonly available paint filter or into another container and label that container "used resin" (not mixing it back in with the unused resin). Resin keeps for a long time, but it is recommended to use up the resin that has already been exposed to light before opening a brand new resin bottle.

- Unless you plan on storing the resin in the vat (which may get dust in it or spill if knocked over), you should wipe the resin vat clean with a small non-abrasive wiping tool.

- You will need to replace the vat over time. The vat that holds the resin is considered a consumable, with a typical lifespan of about 3–4 liters of resin. Vats range in price but average around $50–$80 each.

- mSLA printers (described below) also have a consumable LCD screen which blocks the light to create the patterns to be printed. Over time, the heat from the lamps degrades the screen to be the point of needing a $100-$600 replacement (depending on LCD/printer size). The lifespan on our own $500 resin printer is about 400 hours of screen-on time, resulting in about a 6 month replacement cycle.

Third-Party Resins

While most SLA printer manufacturers allow the use of third-party resin, other manufacturers state in their terms and conditions that their printer warranty will be void if third-party resin is used. We recommend you read the terms and conditions of any printer you want to buy thoroughly to make sure you can use third-party resins.

As you can see, there are a lot of considerations you need to take into account with a resin printer. If you are willing to conduct the process, however, you can enjoy truly exceptional prints, as shown in **Figure 6-2** and **Figure 6-3**.

FIGURE 6-2: Eiffel Tower model printed on an SLA printer (3D model credit: Pranav Panchal)

FIGURE 6-3: lose-up photo of Eiffel Tower with each individual railing measuring just .2 mm

COST OF MATERIALS

Resins are also relatively more expensive than FDM filaments, though they are slowly coming down in price. The least expensive resin can go for $25 per liter (1000 mL) and the most expensive general-purpose resin can go for $100 per liter (these are usually for jewelry burnout /casting processes).

In general, the price of 3D printing with a $30 liter of resin is about on par with printing in PLA on an FDM printer when measured per cubic centimeter. The more expensive resins are similar to printing with specialty FDM filaments (like impact modified PLA, or flexible filaments). A full description of the costs associated with 3D printing in resin can be found in Chapter 9.

How Castable Resins Are Used

Most castable resins need to go through a burn-out process. The resin print is placed in an investment-casting material (plaster) that is hardened, and then the mold is placed in a kiln to burn out the resin inside. This leaves a hollow cavity, and the molten metal used for casting is then poured into that. Castable resins are predominantly used in jewelry design and production.

TYPES OF RESINS

Because of the way resin is formulated, there is less choice in material properties for the SLA printers than what you will find in FDM printers.

THE MOST COMMON TYPES OF RESINS FALL INTO THESE FOUR CATEGORIES:

- Normal/general-use
- Hard/durable
- Flexible
- Castable

There are almost always trade-offs and benefits when moving from one type of resin to another. For example, resins that are made for strength and durability cannot print in fine details like normal, general-use resins. But, those more durable resins can take more physical stress than the general-use resins Some 3D printer manufacturers also manufacture their own resin, and they (usually) strongly suggest that only their specific brand of resin is to be used in their printers.

HERE IS A LIST OF POPULAR RESIN VENDORS.

- Formlabs Resin
- MadeSolid Resin
- B9 Resins
- MakerJuice Resin
- Spot-A Materials
- Elegoo
- SirayaTech

THE THREE TYPES OF RESIN PRINTERS: LASER, DLP, AND MASKED SLA

Laser-based SLA printers trace out the curing path of the resin with a laser. DLP (digital light processing printers project the black and white shape of the layer directly onto the build plate (where "white" gets cured and "black" is not cured). mSLA (masked SLA) printers have a uniform light source on the bottom, and an LCD screen between the light and the vat turns either black (opaque) or "off" to allow light in specific shapes through, one layer at a time. The following chart explains the strengths and drawbacks of each type of printer.

As a general statement, resin printers, per volume of printable area, are more expensive than FDM printers. Entry level FDM printers exist at $300, and resin printers also start at the $300 range. The build envelope of FDM is somewhere around 3x that of resin at that price range, however! Larger size resin printers only really start to come to the market as solid machines around $1,200 and go up to around $3500.

LASER SLA PRINTER	DLP SLA PRINTER	MSLA PRINTER
Very fast for small objects because the laser only has to trace a small area before moving the build plate up.	Similar speed for all 3D prints, regardless of object size, since the entire layer is either cured or not cured all at once.	Same as DLP in terms of speed for both small and large objects
Slower for large objects because the laser has to trace large areas.	Much faster, comparatively, for large objects because the entire layer is cured all at once.	Same base speed as DLP. Some new mSLA LCD's are "monochrome" allowing for 5x faster print times than even previous generation mSLA and DLP printers!
Universally come with proprietary software for printing because it is difficult to control the laser accurately by third parties.	Comes either with proprietary software or a limited selection of free/open source software, depending on the manufacturer.	Generic software is almost the norm, allowing for several slicing programs to be used.
Laser and galvo-controlled mirrors last quite a long time.	An off-the-shelf projector you would use to give business presentations points upwards in this printer. Bulbs have a known lifespan and cost, and these systems are usually larger to fit the projectors and dissipate heat.	mSLA printers are more simple with a light source pointing up to/through an LCD screen. These printers are therefore smaller, and require LCD replacements once degraded from the heat produced by the bulb.
Example: Formlabs Form 3 printer.	Example: Kudo3D Titan 3 printer	Example: Epax3D X1, Elegoo Mars, Anycubic Photon

PRINTER PROFILE: AUTODESK EMBER (DLP TECHNOLOGY)

In 2015, Autodesk released the Ember resin printer at a cost of around $7,500 USD (for a complete kit with extra resin vats, cleaning supplies and starter liters of resin). The Ember printer was discontinued in the first quarter of 2017, but it still serves as a great example of both a resin printer, as well as an open-source mindset existing in the 3D printing community. Ember as shown in **Figure 6-4** is a custom-designed DLP printer and had a very small build envelope of just 64mm x 40mm x 134mm. This small build envelope was ideal for small items like jewelry prototypes and biomedical applications.

Though this printer had a smaller build envelope (build volume/maximum print volume), traditional professional-level 3D printers catering to jewelry artists can cost around fifty-thousand dollars ($50,000 USD). With a $7,500 USD price tag, this printer offered an attractive alternative (see **Figure 6-5**) compared to the previous more expensive options that provided this level of very fine detail.

FIGURE 6-4 (LEFT): The Autodesk Ember DLP printer **FIGURE 6-5 (RIGHT):** 3D Print from Ember printer showcasing extremely detailed hair-like structures (image by Scott Grunewald, 3dprintingindustry.com)

PRINTER PROFILE: FORMLABS FORM 3 (LASER TECHNOLOGY)

On the other side of the resin spectrum is the FormLabs Form 3 printer shown in **Figure 6-6**. This company also started as a successful Kickstarter and has since brought out multiple iterations of its original laser-based SLA printer.

This printer is used by professionals around the world for rapid prototyping and is a reliable and well known printer. As of this writing it is priced at around $3,500 USD. FormLabs creates its own software for use with their printers which allows them to control the print workflow more precisely. In many professional

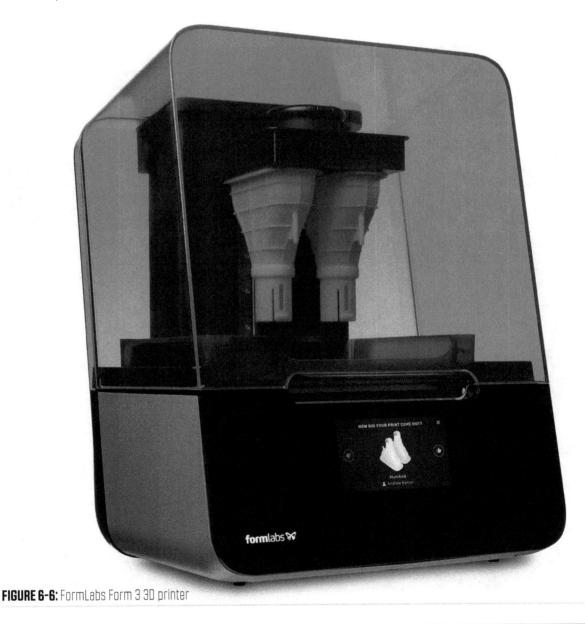

FIGURE 6-6: FormLabs Form 3 3D printer

settings, the slightly increased cost of the FormLabs printer, and FormLabs resin, is well received due to the consistent prints used for rapid prototyping and end-consumer build quality. FormLabs also has some cool technologies built into the Form 3 printer, for example, like the ability to lower the amount of force each layer is subject to upon peeling. This will further positively affect print quality and speed.

The laser-based system shown in **Figure 6-7** is extremely accurate in detail (also see **Figure 6-8**). While it might not be as fast as DLP printers when printing large objects, in professional settings it's excellent at creating consistent prints that require a less manual process.

FIGURE 6-7: Laser tracing an object on the Form2 printer (photo courtesy John Biggs, TechCrunch)

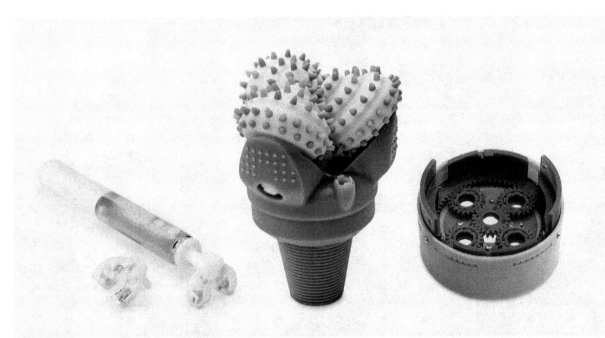

FIGURE 6-8: Sample prints from a FormLabs printer. These objects were printed separately and then hand-assembled.

SOFTWARE: SLICERS FOR RESIN PRINTING

Because the printing process using SLA printers can be affected by many factors, many printer manufacturers have created their own software to manage the print process. Typically, that software is tuned to print with a specific formulation of resin, and if you use third-party resin, you may run into issues. This is especially true with all laser-based resin printers, because the software understands how to move the laser in different directions to cure the layers calibrated to their resin.

Some proprietary software will allow you to tweak settings like exposure time or layer separation time, allowing for other resins to be used, but some software slicers do not open those capabilities to you. As mentioned before, check with the specific printer manufacturer you are researching to see how well it welcomes experimentation from its printer owners.

Conversely, having an all-in-one software system for resin printing can be a benefit. The FormLabs printers use proprietary software called "PreForm" that functions as a slicer for their resin printers. This software makes the preparation of models straightforward and does a great job at creating support structures for the prints while offering convenient presets for FormLabs resins.

Slicers for other resin printers work a little more simply than having to create pathways for lasers to trace. As you read earlier, these types of 3D printers shine a pattern on the build plate. Technically these patterns are just a series of black and white images. For DLP printers, the white areas are the places where the bright light is shining. For mSLA printers, the "white" color is where the LCD is disabled, and the black color

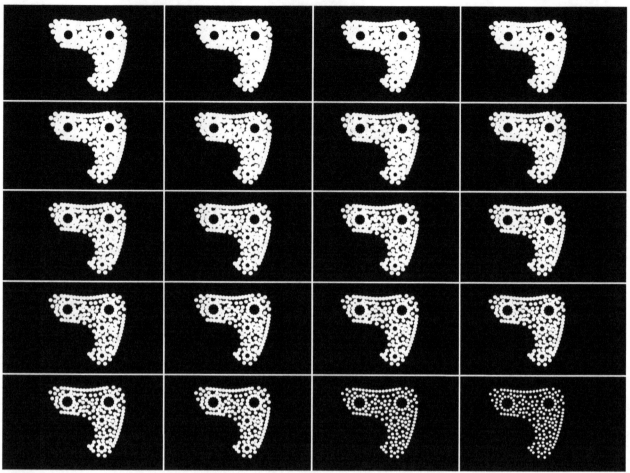

FIGURE 6-9: A succession of layers that would be projected, one by one, on the build plate to create a 3D object

is where the LCD is to turn opaque. In **Figure 6-9** you see a typical set of layers that would be projected on the build plate, one at a time, to form an object.

THERE ARE BOTH FREE AND PAID GOOD SLICING PROGRAMS FOR DLP AND MSLA PRINTERS, WHILE ALL SLA PRINTERS COME WITH THEIR OWN SOFTWARE. A LIST OF POPULAR OPTIONS:

- Lychee Slicer (free and paid): https://mango3d.io/lychee-slicer-3-for-sla-3d-printers/
- Chitubox (free): https://www.chitubox.com/
- Prusa slicer (free): https://www.prusa3d.com/prusaslicer/
  (A note on PrusaSlicer...this software works with Prusa FDM printers, and the Prusa SL1, an mSLA printer. Many people like the support structures created by PrusaSlicer and create the 3D models in PrusaSlicer, and then export the 3D model (including support structures) then use one of the options above to do the actual slicing)
- Slic3r (free): https://slic3r.org/

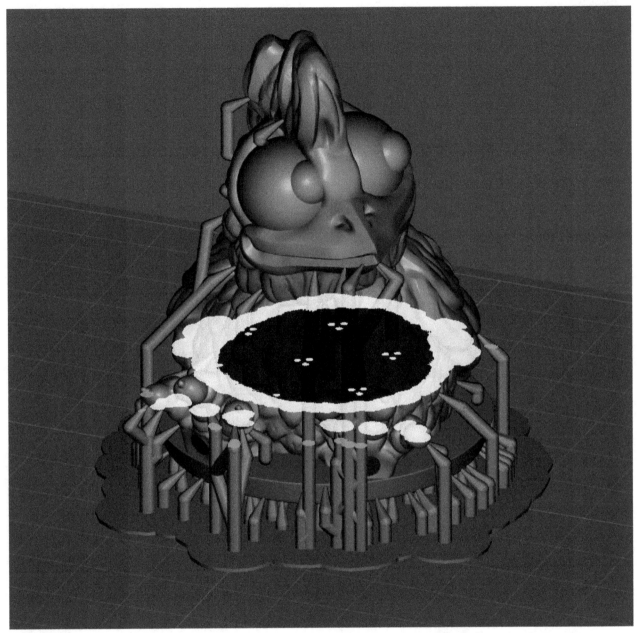

FIGURE 6-10: A model in Lychee slicer with the cross section view turned on, hollowing enabled, and a 5% infill added.

Each one of these slicers takes the 3D model, adds supports to the model, for areas that need it, and then outputs a machine-specific file that gets loaded into the printer. Sometimes these printers exist on the network, but more often they get their printing orders from an SD-card that has the machine-specific file copied onto it. **Figure 6-10** shows a screenshot from Lychee Slicer with a nice little sitting chicken model on the build plate. This model has been hollowed and has small lattice supports on the inside of the model to help with structural rigidity. If you look closely, you can also see small grey drain holes have been added to each foot. In the final print, these holes would also be printed, glued back into the model using some more resin which is then cured, and then sanded to be seamlessly unnoticeable.

SUPPORT STRUCTURES FOR SLA PRINTING

Most SLA printers slowly pull an object, upside down, out of a vat of liquid resin. There are differences that need to be taken into consideration for the creation of support structures using this technology versus FDM technology.

THE THREE MAIN DIFFERENCES ARE:

- The model is being printed upside down, so the effects of gravity on your model's overhangs (compared to FDM) will be reversed (**Figure 6-11**).

- SLA printers can print much more accurately than FDM printers, and so the support structures tend to be much more delicate and thin.

- All SLA prints experience adhesion between the newly cured layer touching the bottom of the vat and the build plate, or previously printed layers. In between actual printing, the printer will "rip" the print off of the bottom of the vat by moving upward, and then move downward again for the next layer to be cured. Figure 6-11 shows the difference between SLA and PLA support structures, including the first layer of supports seen in SLA prints.

- Prints for resin printers want to reduce, as much as possible, the "cross section" of any given layer. A thicker cross section rips off of the plate with more force, and can introduce print irregularities. Printing many models at a 45 degree angle is often performed to reduce that cross section (but this is very dependent on the model being printed).

- You will want to hollow very thick models to both save on resin cost, as well as to reduce the cross section of your printed model.

- If a model has a large hollow area, and with thin walls, the "suction force" created by the build plate "ripping" the newly cured layer off will stress those thin walls as that hollow grows. Think of trying to drink a milkshake through a small straw...it is very difficult, and the walls of the straw collapse inwards...the same thing happens with resin prints. Therefore, long/large hollows usually have drain holes added near, and perpendicular to, the build plate to allow an "air gap" to exist, eliminating suction pressure.

RESIN PRINTERS CONCLUSION

Resin printers require more discipline in order to get to a final print, but for many the process is worth it. Certainly, for those printing with castable resins in the jewelry industry, or who are printing gaming miniatures,

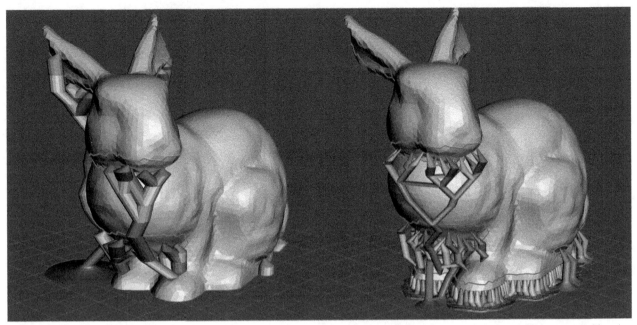

FIGURE 6-11: Differences in support structures generated in Meshmixer with the same overhang settings for an FDM printer (left) and SLA printer (right)

the benefits of a resin printer cannot be overstated. Due to the level of detail, SLA prints are immediately more acceptable as finished products for most people.

You've now learned about FDM and SLA printers. The next chapter discusses the pros and cons of owning a 3D printer versus outsourcing the actual 3D printing to a service.

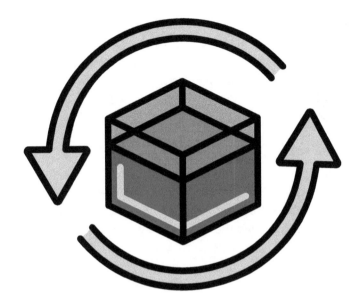

7

OUTSOURCING VERSUS BUYING YOUR OWN 3D PRINTER

In 2016 for the first edition of this book we stated that people were comparing the then-current state of 3D printing to where personal computers were in the 1980s. Now in the second edition of this book we can report....that things have only marginally gotten better. Printing at home is still fraught with problems where your prints may fail for any number of reasons, leaving you with troubleshooting and reprinting. 3D Printer manufacturers have started to build in more intelligence to their hardware and software (Prusa being one such notable company), but 3D printing is nothing like a microwave oven, where you push some buttons and your activity will be successful every time.

After reading the chapters on FDM and SLA technology you might have a preference to either own a 3D printer or to outsource the 3D printing to someone else. If you aren't sure, this chapter will help you decide. Even if you choose to own a 3D printer, this chapter will still be relevant to you. We compare the differences between FDM and SLA, and chances are someday you may want a 3D print in a material that your printer doesn't support, in which case outsourcing the printing may be necessary.

Let's compare the different technologies and options to help you decide what process fits your needs best. We will look at home printing versus outsourced printing and make distinctions between FDM and SLA, as well as consumer versus professional outsourcing (**Figure 7-1**). Please note that these are generalizations. You can receive professionally outsourced prints as quickly as next-day if you are willing to pay double (or more) the printing cost for expedited service. With resin printing, you can mix different ratios of resins together to get the material properties you are seeking, which is a unique feature of that type of 3D printing. The chart in **Figure 7-1** describes the "basic" comparisons of some variables.

	Home Printing		Outsourced Printing	
	FDM	Resin	Consumer	Professional
Cost	$	$$	$$$	$$$$
Speed Of Delivery	Fast	Fast	Slow	Slowest
Material Availability	☑☑☑☑	☑☑	☑☑☑	☑☑
Setup Time	◔	◑	n/a	n/a

FIGURE 7-1: The advantages and disadvantages of home versus outsourced 3D printing

THE BENEFITS OF OUTSOURCING YOUR PRINTING

For many, 3D printing will never happen at home. After all, it is very easy to submit a 3D model online, and have your object arrive on your doorstep 7–10 days later. That object, however, will be printed on a 3D printer that may cost anywhere from $60,000 to $1,000,000.

The graph in **Figure 7-1** shows that outsourcing your 3D printing jobs eliminates the need for printer setup, though at an added cost of production and shipping time. When you send a 3D model away for production, you don't have to set up the printer and monitor the print as it's being produced. You also get the benefit of having more material choices. Speed of delivery is less attractive because you have to wait for your print to be mailed to you, and the cost is higher because you are paying for material and someone's else's time. If you are willing to put up with some failed prints (and the time it takes to restart the print and troubleshoot why the print failed), then you can typically print several iterations of an object in the time it would take for an outsourced vendor to return a single print to you.

But the highest benefit of outsourcing is that it allows consumers to create objects of the highest levels of quality on some of the most expensive machines in the market. The quality of those prints can be near "production quality," making this not only an attractive approach but necessary for some applications.

Many businesses could benefit from custom-created 3D printed products. Promotional giveaways, rapid prototypes of a new product, creation of visual props for sales calls, and many other examples show the value of 3D printing. But not all these companies have in-house CAD designers and 3D printers. They want to participate in the benefits of 3D printing but can't allocate the financial and time resources to bringing on staff for it. That's where outsourcing really has its benefits.

Luckily, there are many options for outsourcing, and it offers an alternative to otherwise costly productions that would be out of the question for many consumers and businesses. For our clients that need 3D printing, they are happy to consult with experts in case they need a specialized material.

CONSUMER 3D PRINTING SERVICE BUREAUS

Consumer-focused 3D printing services like Shapeways, shown in **Figure 7-2**, and Sculpteo, shown in **Figure 7-3** as shown on the next page, offer lower costs than the more professional alternatives by having the quoting process automated, and they also offer a wide range of materials and a more accessible ordering process. Businesses are using these companies to make rapid prototypes and small production runs.

BUSINESS 3D PRINTING SERVICE BUREAUS

If you need specialized techniques or absolute precision in your final model, then moving to a more business-oriented outsourced 3D printing service bureau might be a good option. FathomMFG, headquartered in Oakland, California (**Figure 7-4**) offers not only 3D printing, but production film level prototyping and final part design for high end projects (as well as smaller ones):

- Experts focused on accelerated prototype fabrication & Low- to High-Volume Production Part Services.
- Additive, Traditional, & Hybrid Manufacturing Expertise with an Industry-Leading Advantage of Agility
- 3D Printing, Additive Manufacturing, CNC Machining, Urethane Casting, Tooling, & Injection Molding

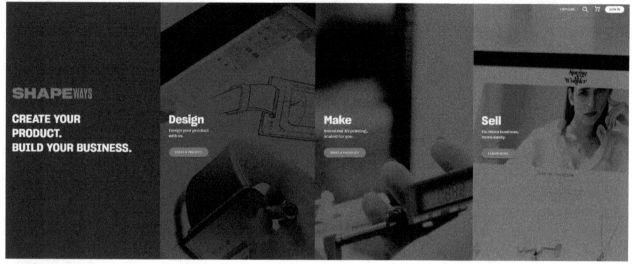

FIGURE 7-2: Shapeways home page

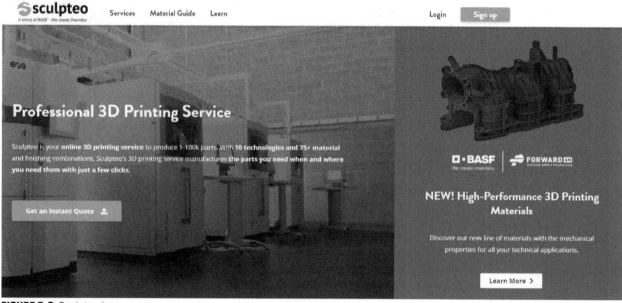

FIGURE 7-3: Sculpteo home page

Another option is Forecast3D, shown in **Figure 7-5**. Just like FathomMFG above, Forecast3D does not have a repository of designs you can purchase like Shapeways and Sculpteo, but it is focused on more high-end fabrication technologies and larger volume production runs for customers with specific mechanical and physical properties requirements.

Forecast3D not only creates high quality individual parts but caters to using additive manufacturing for low-volume production runs. Before additive manufacturing, "low-volume production runs" were meant to

Parts in 1-3 days

Accelerated Manufacturing.

Fathom provides advanced **rapid prototyping** and on-demand **low volume production** services. We are creative problem solvers that deliver high efficiency outcomes. Every time.

SEE OUR CAPABILITIES >

GET YOUR 30 SECOND QUOTE NOW

FIGURE 7-4: FathomMFG home page

FORECAST 3D provides a unique depth of custom manufacturing and 3D printing services to a wide variety of industries including Healthcare, Automotive, Aerospace, Consumer Goods, and Design. For more than 24 years, we have been empowering companies to bring their ideas to life - faster - with the best in Additive Manufacturing technologies.

FIGURE 7-5: Forecast3D home page

be in tens of thousands of units, and there really was not an economical method for runs of a few hundred to a few thousand units. Companies like Forecast3D specialize in high quality smaller runs, as well as offering multi-thousand item manufacturing services that challenge traditional injection molding, in terms of the number of units they can create.

THE RISE OF LOCAL OUTSOURCING

There are also more local options available. A service named 3DHubs has a network of printers around the

US (and some internationally) that are ready to print/create your 3D models. Several years ago, you were able to go on 3DHubs' website and see the names and individual printers of normal people with their own printers, and choose to send your business to specific people (sometimes with printers just in their garage!). 3DHubs was a resource to find local printers to which to send your money to support your local economy. In the intervening years, 3DHubs has abstracted that "personal connection" layer to be more like the other services in this list. But behind the scenes, there are still some hobbyist printers waiting to print your model at the same time as professional manufacturers with $1M+ machines. **Figure 7-6** shows an excerpt from the 3DHubs homepage.

FIGURE 7-6: The home page for 3D Hubs.

3D HUBS WORKS SIMILARLY TO THE OTHER OUTSOURCING SERVICE BUREAUS, BUT WITH SOME NOTABLE DIFFERENCES:

- You may be supporting your local economy by purchasing from 3D printers in your area. 3D Hubs used to tell you who was printing your models, but they no longer do.

- While high-end materials / printing options are available from which to choose, there are still consumer 3D print providers (using printers which cost under $5,000) benefitting from the centralized 3D Hubs website.

- You get access to slightly less expensive materials like PLA as well as ones that are more exotic/experimental like hybrid carbon fiber materials and flexible materials.

Similar to the international service bureaus, 3D Hubs also quotes you an instant price. This might include a flat "setup" fee and the charge to make the print based on the cubic centimeter volume of the 3D model.

If you don't mind searching for a good printing partner at a cheaper cost, then try 3D Hubs. If you are looking for assurances of material quality, or are working under strict business processes and protecting sensitive designs to comply with your own company's internal processes, then purchasing from the other full service providers mentioned earlier may be a better choice for you.

THE BENEFITS OF 3D PRINTING AT HOME

Certainly many of the 3D prints you see on the Internet today were produced in people's garages and homes. 3D printing at home offers the hobbyists and makers faster feedback on their 3D model, as shown in **Figure 7-7**. They can print each version at home, not having to wait days to receive the print as is the case with service bureaus.

Where 3D printing at home costs around four cents per cubic centimeter, it is not uncommon to see a price of 50 cents to $1 per cubic centimeter from an online service bureau (plus a setup fee added on top of that). That represents a 10-25x increase in price over printing it yourself, not counting your own time for setting up, monitoring the print, and cleaning up.

A disadvantage for 3D printing at home is one that is also difficult to quantify. It's the amount of time that it takes you to do everything that "surrounds" the actual 3D printing. This includes leveling the build plate,

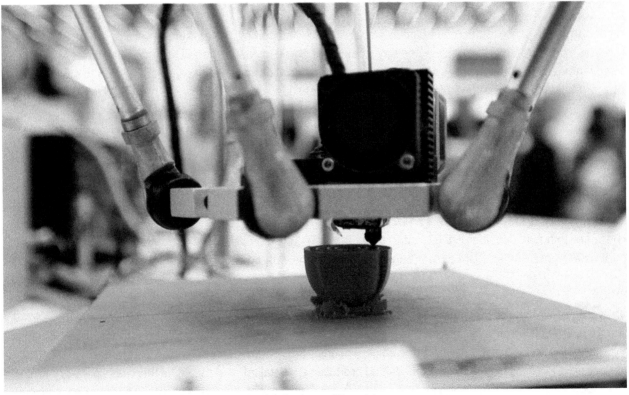

FIGURE 7-7: 3D printing at home gives hobbyists quicker feedback on a 3D model

	FDM	Resin
Printer Cost	$300 to $3500	$350 to $3500
Printable Size	Medium-Large	Small-Large
Print Quality	Decent	High
Print Speed	Medium	Slow to very fast
Materials Cost	$20 to $75/kg	$30 to $150/L
Materials Cost per Unit	2.5¢ to 6¢ / cm^3	3¢ to 20¢ / cm^3
Setup/Cleanup	Relatively easy	Relatively difficult

FIGURE 7-8: Advantages and disadvantages for FDM printers compared to resin/SLA printers for printing at home. The chart includes the general prices for printers in the consumer realm.

making sure your filament is stored properly, and other factors that take time and some knowledge in order to do successfully. For the most part, once you have gone through your initial learning process for 3D printing at home, the process becomes easier. The analogy is like going to a restaurant versus making the food at home from scratch! Although we can say, cooking at home is easier than 3D printing at home! Another disadvantage is not having access to a wide range of material options, unless you are willing to stock many different kinds of materials, in different colors and properties.

But as we mentioned in the beginning of the book, the act and process of 3D printing at home and being a DIY hobbyist is very rewarding. Remember the last time you learned a new skill—woodworking or making ceramics, for instance—and then mastered it enough to get the results you wanted? 3D printing at home provides you with similar opportunities to expand your knowledge and the satisfaction of honing your skills.

VARIABLES INVOLVED WHEN 3D PRINTING AT HOME

There are things to consider if you want to print at home. Let's run down the list of the differences between printing at home with an FDM printer and printing at home with an SLA printer (**Figure 7-8**).

PRICE

On average, FDM printers will cost less than resin printers per cubic centimeter of print area. What this means is that if you are looking for a large print volume, then FDM printers will be less expensive for those larger models. If you are OK printing dental models, tabletop gaming figurines, or other small items, a $300 mSLA (masked SLA) resin printer will be fine for your needs and produce very detailed models. While there are many printers out there (both resin and FDM), pricing for these printers can be described as: starts at "rock bottom price but troublesome" and moves to "acceptable price with low failure rate" as the price ramps up.

At the time of this writing, FDM printers can be had for $300 from companies like Creality (**creality3d.shop**) with printers getting solid with the Prusa MK3S around $750. Small build envelope resin printers can be found from EPAX3D (**epax3d.com**), Elegoo (**elegoo.com**), and Anycubic (**anycubic.com**). These printers are also in the $250-$400 range but have smaller build volumes than the FDM printers above. Larger mSLA resin printers start around the $1200 range.

SIZE
Mainstream FDM printers can print objects up to 11" x 11" x 11" while many entry level resin printers print comfortably at only a smaller 6" x 4" x 6" size.

QUALITY
What the SLA printers lack in volume of print area, they more than make up for in the quality of the resulting print. SLA/resin prints from consumer printers are of such good quality that they can look like injection molded parts. FDM printers, on the other hand, produce visible line layers.

SPEED
FDM printers are limited to an upper limit of about 100–150mm/second print speed, having to trace out each line in each layer individually with molten material. SLA (Laser-based) resin printers work the same way, but they trace with a laser beam and are typically much faster when completing a layer. But one of the drawbacks of resin printing is that the model needs to be separated from the bottom of the resin vat. This process requires force to break away the cured layer of resin and then to move the build plate back down for another curing operation to get the next layer. This puts laser-based systems' speed roughly on par with FDM printers.

Resin printers (DLP and mSLA) cure an entire layer, small or large, at one time. For larger prints, or prints that have a lot of one thing copied over and over (such as jewelry designs), DLP and mSLA printers are significantly faster than laser SLA or FDM printers, and have predictable printing times. For these kinds of printers the print time is equal to the number of seconds the "light" is shining added to the upwards and ownwards movement of the build plate, per layer.

MATERIAL AVAILABILITY
FDM printers offer the most diversity with more than 50 materials (and growing) to choose from. 3D printing service bureaus come in second place with the most variety because they own many different types of printers. Some of these services are even willing to 3D print parts in a castable material and then traditionally cast them using materials such as brass, bronze, or steel.

MATERIALS COST PER UNIT
When you create a model, the space inside of the model can be described using a metric of volume, which is typically "cubic centimeters." Online 3D printing service bureaus will give you prices for materials that are charged at a starting flat rate, and then the number of cubic centimeters of your model is multiplied by the price per cubic centimeter.

If you are printing at home, you have no startup fee and your material cost will be much less. A single spool of PLA filament, 1kg (2.2lb), will represent about 800 cubic centimeters of printable material and costs around $21 to $35. More exotic filaments like PET or flexible filament are roughly double the price of PLA filament, thus in the $70 range.

Resin printing has a similar price per cubic centimeter compared to FDM printing when printing with the lower cost resins. One liter of resin contains about 900 cubic centimeters of printable material. 1Kg of filament is about 800 cubic centimeters. Assuming a parity between 1L of resin and one spool of PETG, the prices are comparable. Resin prints tend to have thicker walls than FDM as there is no time penalty for printing a thicker wall on mSLA printers. In the FDM world, those thicker walls would take quite a bit more time to print. The thicker walls on resin prints serve to make the objects more structurally sound, but at a higher material cost.

TIME NEEDED FOR SETUP/CLEANUP

This is the other critical factor that sways people in one direction or the other. With the online print services, the base startup fee for an FDM print can be around $5. For resin prints, startup fees can be around $20. Why the difference? Resin is much harder to work with than FDM. When you run a resin printer, you have to use gloves to prevent the resin from getting on your hands, as well as pour the unused resin back into containers for long-term storage. A resin print also needs to be washed off with isopropyl alcohol to remove the uncured resin, and it needs some time in an ultraviolet curing box (or to be placed in direct sunshine) to finish curing.

For an FDM print, you just pop the print off of the build plate and everything is pretty much done. Of course, if your workflow requires extreme detail, or you do not mind working with resin, you will enjoy the prints that resin printers create.

PRODUCTION QUANTITY: LOW-VOLUME MANUFACTURING

If you are looking to create many units of a design, you have three options: 3D print them at home, go through a service bureau, or use traditional manufacturing. Typically, creating a mold for injection molding (traditional manufacturing) costs anywhere from $2,000 to $10,000 for a simple mold, and then the prices increase for molds that are more complex or that are made for producing many thousands of items. Before 3D printing there was a gap in the market where inventors and product designers had very few inexpensive options if they wanted to produce a thousand or less copies of an item. The economics for traditional manufacturing (with molds) only made financial sense when production neared the "several thousands" unit range.

Additive manufacturing has changed all of that. Prototypers can move to smaller production runs at significant fractions of the cost that previously were commanded by manufacturers. The economics of making many thousands of items still favor traditional manufacturing, but a new class of "boutique" manufactures is being created by 3D printing.

CHOOSING THE RIGHT QUALITY FINISH FOR PRINTS

If you want to sell products to consumers and the quality of surface finish is important, you can outsource your printing to a 3D printing service bureau to get excellent quality at a higher cost per unit. Or, you can print the objects yourself.

If you decide to print the objects yourself, you have several options for surface finishes. If you are trying to compete with traditional parts (that have smooth surface finish), then any FDM print will not be the right choice without what is called "post processing."

This is a generalized list that applies to both FDM as well as to resin prints. Post processing steps can also apply to prints made at service bureaus, even from their very expensive 3D printed titanium parts!

SOME POST PROCESSING OPTIONS MIGHT INCLUDE:

- Surface sanding (either manually, or mechanically) using smoothing machines and sanding medium like glass beads or plastic pellets

- Milling operations to create perfectly dimensioned holes for screws

- Tapping (the act of cutting threads into an object rather than 3D printing the threads)

- Heat set inserts as shown in **Figure 7-9**. If you want excellent threads to accept screws, you can heath these up with a torch and press these into the 3D printed plastics to lock them in and provide great threads

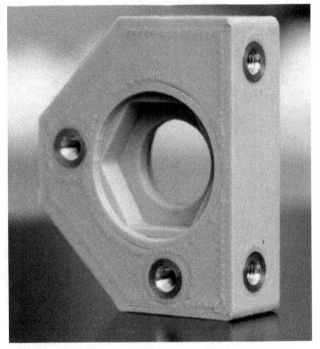

FIGURE 7-9: : A heat set insert which would get pressed into a 3D print the enable reusable threads (Image credit: Joshua Vasquez, via Hackday article https://hackaday.com/2019/02/28/threading-3d-printed-parts-how-to-use-heat-set-inserts/)

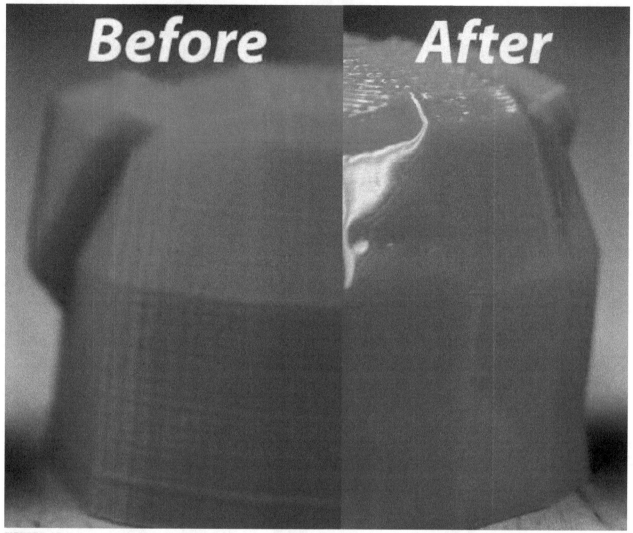

FIGURE 7-10: Before and after results of treating the outside of an ABS print with a few seconds of acetone vapor

If you decide to use FDM printing for creating end-use parts you can use a material called ABS (acrylonitrile butadiene styrene). As previously discussed, ABS is a durable material but can be a bit smelly to print with; it also requires a heated print bed. ABS has one unique capability, however. The chemical acetone "liquifies" ABS. You can use acetone vapors to "melt" the outside layers of your print, as shown in **Figure 7-10**, to create a smooth finish.

Acetone: Proceed with Caution

Acetone is an aggressive chemical that is not recommended for inhaling or direct skin contact. The acetone vapor you will be producing by heating the acetone is extremely flammable. Exercise caution, otherwise explosions or fires may occur.

The other method for making production parts out of FDM prints is to post-print hand-process them. This could include sanding the prints to remove the obvious layers created by the printing process, as well as painting the model to cover up the remaining layers. It is a laborious process, but one that can yield nice parts, at the expense of your time.

EVALUATING FOR SPECIAL MECHANICAL CONSIDERATIONS

Even if your product does not need to undergo extremes in terms of durability or performance under stress, something as small as a pair of earrings still undergo the forces that are common in life. Objects get dropped, or brush up against a wall, or get tossed in buckets with other items. You need to make sure that your print can withstand the environment in which it will be placed. On average, it's best to start with FDM prints, even if those prints might not have the best surface quality as compared to resin prints.

Another main reason to prefer FDM over resin is the wide variety of FDM materials available to the consumer, which gives you more choice in suiting the print to its end goal.

As an example of the breadth of filament available, **Figure 7-11** shows a spool of impact modified PLA (polylactic acid). It looks like normal filament, but looks can be deceiving!

FIGURE 7-11: A spool of impact modified PLA

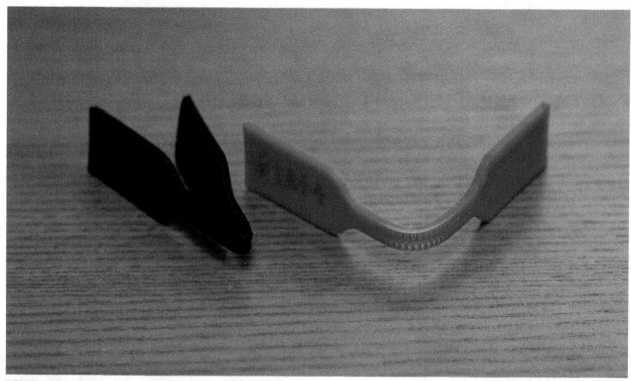

FIGURE 7-12: : After three back-and-forth bends, normal PLA (left) snaps, while the enhanced PLA (right) remains structurally sound

The impact modified filament in **Figure 7-12** is an example of a "blended PLA" that was designed to improve on PLA's characteristic to be rigid, but brittle when exposed to stresses. There is a growing field of "functional filaments" created by a number of reputable companies. They can create not only decorative items, but items that are designed to take advantage of specific material properties.

SOME NOTEWORTHY BRANDS OF "SPECIALIST" FILAMENT MANUFACTURERS ARE:

- NinjaFlex provides flexible filaments, more generically called "TPU" (thermoplastic urethane) filaments

- PETG is similar to what water bottles are made out of and is one of the most popular materials for FDM printing

- Many types of "enhanced" or "impact modified" PLA

- Nylon is very strong but slightly more difficult to print with due to its tendency to absorb water quickly from the air thus necessitating keeping the filament dry between print runs

As an example of what a functional filament is designed to do let's take impact modified PLA. PLA is a great material to make strong, rigid objects. But, once your object gets stressed to a failure point, normal PLA will snap or shatter. Look at the following example prints to see how a normal PLA part (left, in black) fails while the "enhanced" PLA (right, in blue) survives. Each sample was bent backward and forward three times, as shown in **Figure 7-12**.

The rise of enhanced/functional materials that overcome the limitations found in current filaments will increase. We will continue to see advancements in both FDM and SLA materials, making them more adaptable and useful in more applications.

CONCLUSION

You are now more familiar with the options you have when 3D printing your models. You will need to consider your resources in terms of money, time, and willingness to work with technology that sometimes fails when making the decision to print at home or outsource the job to a third party.

It's great fun to watch the printers operate, with their buzzing and whirring, and then seeing the results of your work turn into a physical product. Additionally, for many people, the enjoyment of producing 3D CAD models themselves will be more popular than the actual physical printing process. You won't want to miss the next chapter, which walks you through the whole 3D printing workflow from idea, to CAD model, to print!

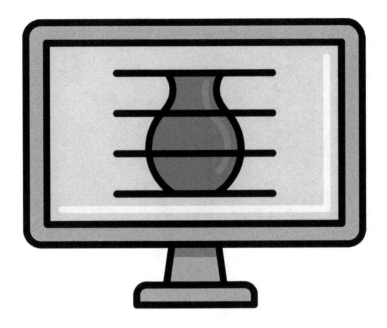

OVERVIEW OF THE 3D PRINTING WORKFLOW

FIGURE 8-1: A 3D modeler working with CAD software

The previous chapters introduced you to 3D printing terms and various 3D printing options. This chapter will help you get your mind around how an idea becomes a 3D print. We will explore the processes and tools that we use in the rapid prototyping part of our business and explain how you can apply them at home.

You may have seen a 3D printer in action at a local technology event or retail store, or via an online video, and decided that you wanted to try it yourself. You might have even signed up for an account and downloaded some free modeling software, as shown in **Figure 8-1**.

And...a couple of hours later you were frustrated with the uncooperative lump of digital clay on your computer screen. You still have positive feelings about 3D printing's potential but have experienced the disappointment of not being able to get the idea "from your mind to a design" on the computer. Don't worry. You wouldn't be the first person to have this experience.

3D printing is entirely within your reach, but you will need to approach it as you would any hobby or new technology and anticipate a learning curve. First, we'll look at the whole process (or workflow) in overview. Later, we'll look in greater detail at how to work successfully with that 3D "modeling clay."

FIGURE 8-2: A still image from Cosmos Laundromat ((CC) Blender Foundation)

3D FILES

Before you can print anything, you have to have a 3D model that "tells" the printer what to print. The most common type of file format in the field of 3D printing is STL (you will see it as .stl), which stands for "STereo-Lithography." Don't be tricked, though; this file format is used by all 3D printers, regardless of type. The STL file simply defines points in space and connects those points together to form a series of triangles, which in total, are called a "mesh" or "mesh object." The file is called a "shell model" and can be thought of as a monochrome, thin skin with a hollow interior.

These digital files are then sent to a software program, called a "slicer," which cuts a digital file into many tiny "slices" that instruct the 3D printer where to trace and lay down each layer to make the 3D object. Imagine taking a loaf of bread and slicing it into hundreds of horizontal slices. No two slices would be exactly the same—each would be slightly different from the ones right next to it. When they are all stacked one on top of the other, they end up re-creating a loaf of bread. If you have ever seen a medical MRI or CT scan, it's a similar process.

3D MODELS

You've seen 3D models all over the place! The latest science fiction and animated movies are filled with 3D models, like this one (**Figure 8-2**) from an open source film titled Cosmos Laundromat, which was created in free modeling software called "Blender" (**blender.org**). Though this movie is around 6 years old, it is still a great example of how great models can be created using free software.

3D models are a digital representation of an object. Many of the 3D models created for movies and games were not modeled with the end goal of 3D printing. If you tried to download those models, the model would probably not be "optimized" for 3D printing and would need significant work in order to print. (We will discuss this in greater detail in Chapter 11, where we go over how to fix a 3D model in Meshmixer.)

Many of the real-world products we use in our daily lives began their existence as 3D models, such as the bike frame modeled in a computer-aided design (CAD) program called Fusion 360 as shown in **Figure 8-3**.

Future File Format

As of this writing, the industry standard is still the STL file format. In April 2015, Microsoft and other prominent 3D printing companies formed a consortium to define an alternative file format that they claim improves on the STL format. It's called 3MF, and Microsoft said that it is designed to make 3D printing easier and more manageable. The new format would reduce loss of detail when exporting files for 3D printing, and it is designed to define vital information like color and material specifications. The 3MF file format has promising potential but has not been standardized in the consumer realm quite yet (even since the first publication of this book!).

3D modeling programs are a great way to express yourself creatively and to bring your ideas into reality, and we will help you get started with them in the next section of this book. If you don't want to make your own, however, other people have created 3D models and have shared them freely online. You may want to start by printing out some of these pre-built models. You'll find a more detailed description of where to find models and how to evaluate them in Chapter 13.

HOW DO I GET FROM IDEA TO OBJECT?

In the rapid prototyping division of our company, HoneyPoint3D™, we tell customers that they can bring in an idea or sketch and we can turn that into a computer-aided design (CAD) file that can later be 3D printed. Many times, what they bring is a drawing on a piece of paper, sometimes even on the back of a napkin! **Figure 8-4** shows the basic process of how we turn an idea into a 3D print.

Some clients come into our office with nothing more than an idea, but generally people have at least 2D drawings from which we can extrapolate measurement details. Keep in mind that it's ok to have "best guess" measurements when you start modeling. As you move through the process of creating models, you will later be able to change and refine the measurements. This rapid-prototyping process (described in Chapter 15) can be applied to creating anything from a princess figurine to a car engine.

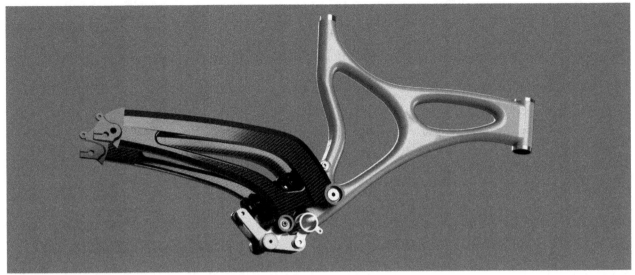

FIGURE 8-3: A 3D model of a bicycle frame created in Autodesk's Fusion 360

FIGURE 8-4: The workflow from idea to finished 3D print going clockwise from the left: 2D drawing, CAD file generation, 3D printed prototype, final piece

DETAILS OF THE 3D PRINTING WORKFLOW

Here's a step-by-step guide for how to turn the idea in your mind into a physical object you can hold in your hand:

IDEA: Come up with an idea for an object you want to 3D print.

SEARCH: Thousands of 3D models that already exist. Search on Thingiverse.com or other online repositories for open source 3D printing communities to see if someone else has already designed it for you. If they have, download the STL file and proceed to the slicing step.

PLAN: If it doesn't already exist as a 3D file, draw your idea out on a piece of paper, with several different views. Draw your object from the top, the side, and the front. If you can, and put in approximate measurements.

3D MODELING: Now that you have a plan, it's time to 3D model it in a software program. If you are a beginner to the CAD modeling world, we recommend you start with Tinkercad (**tinkercad.com**). If you are more experienced or want to create something more complex, then use Autodesk Fusion 360 (**fusion360. autodesk.com**). For organic shapes, use Meshmixer (**meshmixer.com**). Starter tutorials for all three CAD programs are provided in Chapters 10, 11, and 12.

SAVE AND EXPORT: In all of the previously mentioned programs there is an export function that will allow you to save your design in STL format. Save that STL file and download it to your computer.

SLICE: Open the slicing software recommended by your 3D printer manufacturer and load your model. It's best practice to use the program that your printer manufacturer recommends first, and then move to other slicing programs later if you desire. Slicing an object means manipulating the original 3D file into many tiny horizontal slices that go through the entire object, as shown in **Figure 8-5**. These slices will later become the pattern by which the 3D printer knows where to trace and deposit material, or where to solidify the resin.

EVALUATE SUPPORT STRUCTURE: Before you save your print file, you will use your slicer to determine if your model needs support structures to "hold up" overhanging parts that could droop. Support structures are needed to support parts of your model that would droop due to gravity during the build process if they were not supported. Support structures are meant to be a hand-removable scaffolding that the slicer would create for you automatically. See **Figure 8-6** for an illustration of the original model and then the support structures added by the slicer.

CREATE G-CODE/MACHINE-READABLE FILE: The slicer will create a digital file of printer-readable code that will direct the movements of the printer itself. Typically, this is a G-Code file, or it may be a printer-specific file format. It may be named something else depending on the manufacturer of your printer, or it may be a series of .png images if you are using a DLP-based SLA printer. If you open a G-Code file in a text editor, there will be thousands of lines with XY coordinate changes, plus extrusion control commands for the printer to implement.

UPLOAD G-CODE: Upload the G-Code/machine file to the 3D printer from a direct connection on your computer, over a wired or wireless network, or to an SD card that will get inserted into the 3D printer.

MAKE SURE YOUR PRINTER IS READY: As we mentioned in Chapter 5, your print bed needs to be level to the nozzle. If your print bed is not properly calibrated, your print will fail. Most printers do not automatically calibrate themselves, so you will need to pay close attention to the manufacturer's instructions on how to

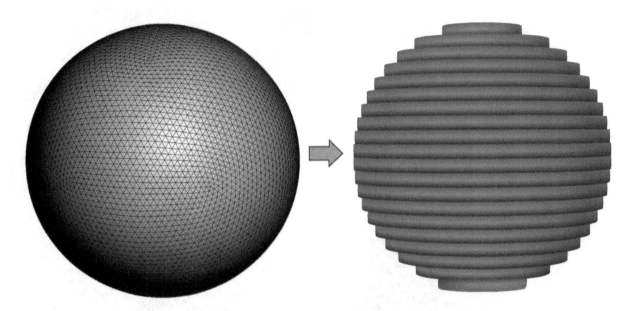

FIGURE 8-5: Example of an object being sliced into many layers in order to get it "print ready." Each blue layer would be printed on top of the underlying layer.

level the printing bed. Manufacturer's instructions will also include how to properly load and store the materials you will be printing with and monitor the print while it's in operation.

PRINT: The big moment! Hit the print button and watch the magic happen!

VERIFY: Well, to be clear, it's not really magic that is happening. Make sure the first layer goes down evenly and well, and keep an eye on your printer throughout the build process. No 3D printer should be left unattended during the print process, and most 3D printer manufacturers will state this in their instructions.

POSTPROCESS: After the print is complete, remove it from the build plate by hand or with a small prying tool, and remove any support structures by hand or with pliers. If you are using a resin printer, you will need to wash/clean your print and post-cure it for a few

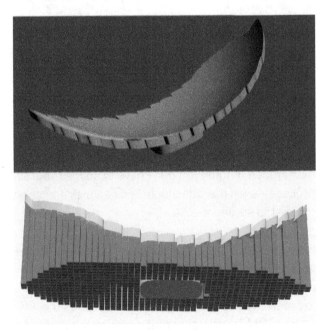

FIGURE 8-6: (Top) A curved leaf on a pedestal. (Bottom) The curvature of the yellow leaf with automatic supports added in the slicer to make the overhangs printable. The supports (shown in orange) will later be removed by hand.

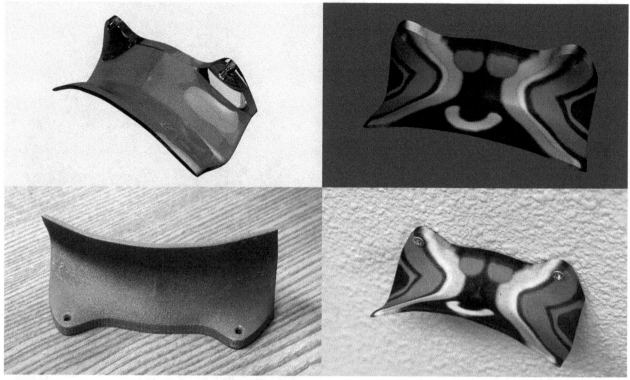

FIGURE 8-7: A wall-mountable headphone holder. (Top left) A CAD model in Fusion 360. (Top right) Model colored in Meshmixer. (Bottom left) Model printed at home using PLA material. (Bottom right) The color 3D printed version made by Shapeways.com.

minutes in sunlight or under a UV curing light. The workflow outlined here might seem daunting at first, but many of the steps do not take that long and will quickly become second nature to you. You'll feel a great deal of creativity and satisfaction once you have mastered the processes and begun making your own 3D prints!

YOU CAN OUTSOURCE ANY PART OF THE 3D PRINTING WORKFLOW

Remember, though, you don't have to do it all yourself. You can outsource any part of the 3D printing process—even the idea stage! You can pick and choose which part of the workflow you want to do and leave the rest to someone else.

In the previous chapter we outlined the difference between 3D printing at home or outsourcing the job to a service bureau like **Shapeways.com**. If the process of 3D printing seems too much for you, outsourcing the printing part of the workflow is very easy and produces great results. (You might have to wait 10–14 days to receive your prints.) **Figure 8-7** illustrates the difference between a 3D print made at home using a con-sumer FDM 3D printer versus using an outsourced service like Shapeways.com.

You have learned the steps needed to make a 3D print on your printer. The next chapter describes in more specifics where to find models, the types of 3D models, and considerations you can take into account when creating your own. Turn the page!

9

SETTING UP YOUR PERSONAL MAKERSPACE FOR 3D PRINTING

FIGURE 9-1: FabLab: The Science Dissemination Unit (SDU) in Trieste, Italy (photo by: Moreno Soppelsa)

Whether you call it a makerspace, FabLab, hackerspace or community workshop, the names all refer to community spaces where creativity and exploration are valued over "getting it right the first time." These community-operated physical spaces are where regular people with common interests can meet, socialize, and make.

First and foremost, the people who use makerspaces share the desire to experiment, tinker, invent, and learn. The actual tools that exist in a makerspace come second to that inventive drive, and in fact, many makerspaces start out with nothing more than a group of people coming together to help each other learn. The more fortunate makerspaces will have 3D printers, laser cutters, CNC machines, electronics components, and various hand tools, as shown in **Figure 9-1**.

The goal of this chapter is to help you create much the same feeling and functionality in your home as you would find at a public makerspace. Creating the perfect makerspace is not about how many tools and pieces of equipment you have. Rather, it is about making sure you are able to be creative and productive.

So, don't feel as if you need to spend a lot of money on fancy tools or have the perfectly organized space (as shown in **Figure 9-2**) in order to make a space you can create in!

In reality, your makerspace likely won't be as neat as in the above image. Even with the best of intentions, your makerspace will more likely end up looking like a mad scientist lives there (**Figure 9-3**) and that's actually a good sign!

FIGURE 9-3: What a real 3D printing maker's workstation looks like

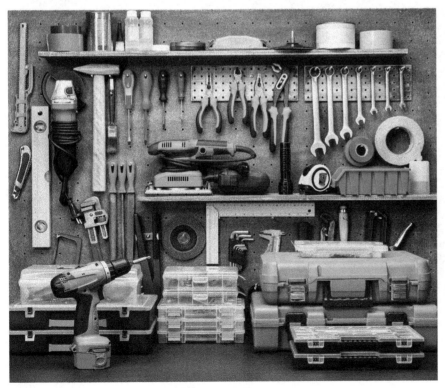

FIGURE 9-2: The perfectly organized mythical makerspace

FIGURE 9-4: Mendel Max 2.0 assembled kit (makerstoolworks.com)

GETTING READY

Here are three essential tasks we recommend you do to get ready to start creating your 3D printing workspace, along with advice on how to best achieve your goal. You'll find it's really helpful to have your space well set up before you begin! We've provided a checklist at the end of this chapter so you can make sure you have the essential tools.

TASK 1: RESEARCH WHICH 3D PRINTER TO BUY

You don't need a 3D printer to participate in this technology, but a 3D printing space wouldn't be the same without... well... a 3D printer. In Chapters 4 and 6 of this book, we identified some 3D printer qualities to look for when purchasing a 3D printer, but this section will give you more insights.

At last count, there were more than 200+ 3D printer manufacturers in the consumer market. They range from startups to multimillion-dollar companies. To get a real sense of the manufacturer's quality, we recommend you visit their forums and read the comments other owners have made.

Decide how much or how little time you want to spend building and maintaining the printer yourself. We put together our first 3D printer, the Mendel Max 2.0, as shown in **Figure 9-4**, on our dining room table. It took 16 hours over the course of 3 days. We learned a lot from the experience, but, if you don't have the time or patience, we recommend you buy one already assembled. Pre-assembled 3D printers are more expensive but worth the money if you would rather focus on 3D CAD modeling or 3D printing rather than learning all the "ins and outs" of how a 3D printer works. Whether you buy it assembled or not, you will still have to learn how to maintain the 3D printer.

You may also want to take a look at Make: Magazine's annual issue devoted to 3D printers or visit their companion website. The magazine's editors select dozens of the latest 3D printers to rigorously test and score.

TASK 2: CREATE A SAFE WORK AREA

Whether you decide you want your makerspace to be in your garage, your home office, or your shed, you will want to make sure you have a safe environment and one that is conducive to your well-being.

You want to be able to move around the workspace easily, have good lighting, tape down your various power cords so you don't trip over them, and have a fire extinguisher within 6 feet. Just in case.

In general, 3D printers are safe machines, but you will want to take precautions to make them safe for other members of your household, including spouses, kids, pets, visiting relatives, and curious neighbors. Certain components of the 3D printer (like any machine) need to have some "look only" areas explained in order to keep everyone safe.

One example of a "look only" area is the extruder (or hot end). You should take caution to not allow anyone to touch the extruder assembly while it's in operation, as shown in **Figure 9-5**. The nozzle can reach hundreds of degrees in temperature.

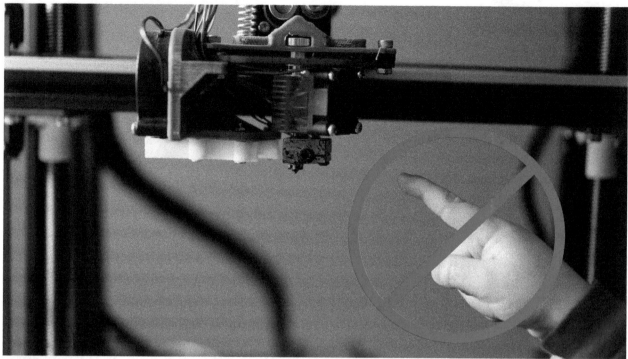

FIGURE 9-5: Some 3D printers have easily accessible heating elements. Avoid touching the extruder assembly when the 3D printer is in operation. (We turned it off to take this picture!)

There are other components you need to watch. Many 3D printers have belts and rods that move, which can present pinch hazards. Exercise caution and make sure to keep fingers, hair, and clothing away from the moving parts when they are in operation (**Figure 9-6**).

Many 3D printer kits have exposed electrical wires and non-enclosed electrical components. 3D printers also run on electrical power from a wall outlet, so normal wall-voltage electrical appliance caution is warranted.

When you are removing a print from the print bed with a sharp spatula, watch your fingers to avoid getting cut. When removing support material from a 3D print, wear protective glasses and make sure your peripheral tools (picks, rotary sanding tools) are stored in a safe place. There are more safety tips you will learn from user forums, the manufacturer's website, and experience.

TASK 3: CHOOSE THE BEST PHYSICAL ENVIRONMENT FOR YOUR 3D PRINTER

One of the nice things about creating a makerspace for your 3D printing work area is that the actual equipment, tools, and accessories can be contained in a relatively small area. Many people even use 3D printers directly on the same desk as their computer! We recommend, however, that you use a table that is dedicated to the 3D printer, as we learned the hard way, movements, spilled coffee, and bumping the desk are not conducive to good prints.

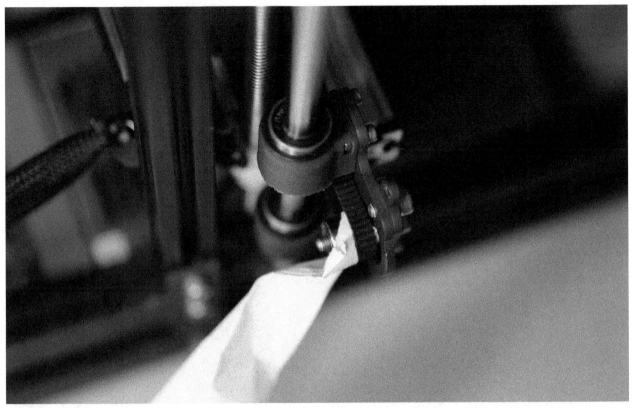

FIGURE 9-6: 3D printer belt and pulley crushing paper, showing the potential pinch hazards

In the beginning of our business, we had more failed 3D prints than we care to remember. It often took hours of trial and error to figure out what factors actually caused the failure. You can get a jump ahead by learning from our experience!

HERE ARE OUR TOP ITEMS TO CONSIDER WHEN CHOOSING THE BEST ENVIRONMENT FOR YOUR PRINTER:

DOES THE ROOM TEMPERATURE FLUCTUATE TOO MUCH? Garages are notorious for heat and cold fluctuations. In the summer they are too hot, in the winter too cold. Too-high temperatures can negatively affect the viscosity of resin, as well as causing stepper motors to not work properly and potentially overheat. Find a place where you can control the temperature and keep it around 70 degrees Fahrenheit. We print in our garage, so we bought a powerful heating and cooling air conditioning unit to moderate the temperatures. Check the BTU ratings before you buy, and make sure the ratings match the size of your room.

GOT KIDS? We have a very curious eight-year-old who loves to get his hands on anything that looks even vaguely interesting.....and 3D printers certainly fit that description! Assume little hands will want to touch the sound-making, toy-producing machine and place your 3D printer appropriately out of their reach. In addition to their safety, any stray touch can upset the calibration of the printer bed and cause prints to fail.

Third-Party Resins

Although many say that the fumes from heating filament in a 3D printer are safe, we feel it's prudent to run a HEPA filter near your 3D printer and/or to have some sort of ventilation near an open window that isn't blowing directly on the 3D printer. The filament material ABS, for example, smells much stronger than PLA for a reason. The "S" in ABS stands for "styrene," which is a known irritant to the respiratory system.

If you want to learn more, take a look at a study done by the University of Illinois that tested 3D printers for fumes emitted by FDM printers. The study found that 3D printers while in operation produced nanoparticle emissions similar to or higher than cooking with a gas stove. These nanoparticles can be absorbed more easily into the respiratory system than other types of particles so it is important to provide for the best ventilation you can. The study did not say what effects those nanoparticles would have on a typical person,

FIGURE 9-7: An enclosure well suited for the removal of fumes / odors caused by the printing process. Note the exhaust duct coming from the top, venting to outside the building.

but the study did state that ABS produced 10x the emissions of PLA—which is why we recommend printing with PLA. As an overarching statement we will say this: any time you are printing with any type of printer, use ventilation. This applies with odorous filaments, as well as ones you cannot smell. Assume there are fine particles coming from the printer in the air, and figure out a way to get good airflow in your fabrication area.

Figure 9-7 shows a relatively inexpensive ($250) setup that helps remove odors and particulates from the print area. The setup is housing two EPAX3D resin printers, but would work well for FDM printers as well. This particular enclosure was actually designed for growing marijuana, but was purchased for a different but associated (reduce odor) use. If you look at the top of the picture, you can see a metallic duct connected to a fan on the inside top of the enclosure. This duct vents outside, with odors propelled by the fan. This design works very well for removing odors, but also carries another risk: If any 3D printer inside were to have a "thermal event", otherwise known as a fire, the enclosure may help the flames spread. Keep in mind the previous advice to always monitor your 3D printer(s) when you are running prints!

BEWARE OF AIRFLOW AND SUNLIGHT. An errant breeze from an air conditioning unit that you cannot feel unless your hand is behind the printer, can spell disaster on your printing process, cooling some sections of your print but not others. An open window can bring some unwanted heat or air currents as well. You should still have proper ventilation, just be sure to place the printer not in line with air currents. If you're printing with resin, you will want your print station away from sources of UV light in order to not prematurely cure your resin.

BEWARE OF UNWANTED MOVEMENT. Whether you have a high-traffic area or an unstable table that wobbles too much when the 3D printer head is moving back and forth, realize that every bump and shake of the table can affect your printer in unwanted ways. Earlier we mentioned that it's not a good idea to have your 3D printer on the same desk as your computer. This is why.

THINK ABOUT POWER SOURCES. Your workstation should be within good distance of a power outlet that you can use with surge protectors. People even put UPS (uninterruptible power supply) systems on their printer's electrical line so if the power goes out, the print will not be lost. Some printers (like the Prusa MK3S) have a feature where if the power goes out, then next time the power comes back on, the printer will warm up, and start printing from the last place it left off. Only a small surface defect will be evident if this happens (hopefully!).

TABLE PLACEMENT. It's preferred that your work table not be placed against a wall so that you can move around all sides freely. 3D printers need maintenance, and some have components located on the back that you might need to access frequently.

3D PRINTERS MAKE NOISE. Some models can take hours or days to print. We recommend not placing the printer in your bedroom, or even the shared wall of your bedroom.
If your printer does not come with the ability to self-manage its own prints via SD card, you will need to dedicate a computer to the printing process for the entirety of the print. Make sure your computer does not enter into sleep mode during the print!

STORE YOUR MATERIALS PROPERLY! We talked about proper storage and handling of both filament and resin in Chapters 4 and 6 respectively, but it's worth stating again. Many filaments absorb moisture from the air, and keeping that moisture out will be one key to successful prints. Though resin does not have moisture issues, temperature changes can cause failed prints when working with resin.

SETTING UP SHOP IN YOUR GARAGE? Here are some special considerations and suggestions to keep in mind:

- Dust is not your friend. Keeping the garage area, floor, and workspace free of dust will help prevent dust from being an unwanted part of your print!

- As mentioned before, try to get the temperature to stay as consistent as possible. We have a lot of windows in our garage, so using 3M foil tape helped keep the temperature inside more consistent.

- Keep open windows about 6 feet away from your printer to allow for proper ventilation. Or use a HEPA filter that is rated for VOCs. Weather permitting, open the garage door to get good ventilation.

- If possible, the environment in your garage should be relatively dry because too much moisture could render the filament unusable. Filament should remain dry and stored properly in a plastic bin with silica packs, on the floor of the garage.

- If you keep a refrigerator or freezer in your garage, set up your workspace on the opposite side of the garage. We don't recommend that your printer share power outlets with other major appliances like refrigerators, freezers, washing machines and dryers, etc.

- If you are printing with resin, have your disposable gloves and washing station easily reachable to prevent drips of rein on your floor. Also a paper towel stand is useful for wiping up any spilled resin (you can even cure that paper towel to harden the resin for easier disposal). Resin printing requires an extra water source/post-processing area.

- Have at least a 3' x 3' x 3' space for the printer and your work area.

- Get a standard tabletop or workbench. You may need one or two of these depending on the size of your 3D printer. Each should be at least 24" x 24" with a 28" sitting height.

- Locate and use a nearby electrical outlet with an attached surge protector to protect the printer.

- Allocate a space for your computer (if your printer requires a physical connection to one).

- Have handy a standard USB cable to connect the printer to your desktop computer or laptop (if necessary). Your printer may have come with one.

PRINTING AND PROCESSING ITEMS TO CONSIDER BUYING:

1. Have one of the following to help FDM prints stick to the build plate: extra-strength glue stick, blue painter's tape (try to find 2" wide or wider if possible), or for those of you printing with a heated build plate, Kapton tape. If using painter's tape, avoid brands with excessively waxy residue that might prevent adhesion.

2. Get a 3D printer toolkit (for edging and scraping excess glue off the print bed after printing). Octave 3D Printer Tool Kit A is a good choice for around $20.

3. Buy a thin spatula (to help remove prints from the build plate). We like using a thin cookie spatula, sometimes marketed as a scrapbooking tool in arts & crafts stores.

4. Consider a small blowtorch for resurfacing and finishing print jobs.

5. Buy flush-cut wire cutters.

6. Buy wide angle and needle nose pliers (for support structure removal, as well as for removing parts from the build plate).

THINGS TO BUY FOR CLEANING UP AFTERWARDS:

$ A small wastebasket for discarded pieces of the print or small pieces of filament you cut off when changing filament. Or, for a failed print.

$ Paper towels (and water) to clean the build plate if using the glue stick.

$ Clean cloth to remove excess glue from the print bed after completing a print.

$ A wire brush to clean the extruder's toothed gear if the teeth get jammed with filament, or if the nozzle gets small pieces of debris on it.

$ .3mm (.012") guitar string to feed back up through your nozzle if you get a clog.

FOR RESIN PRINTERS YOU WILL NEED A DIFFERENT SET OF CLEANING TOOLS:

$ Paper towels

$ Disposable gloves

$ A hard metal scraper (and if you are using extra hard resin, a hammer to knock the printer free using the scraper's end)

$ A soft plastic or silicone scraper to clean the vat on resin switchouts

$ Disposable paint filters, or a reusable cooking filter to filter out small cured bits in the vat if/ when a print failure happens

$ A container filled with 90% (or higher) isopropyl alcohol, or denatured alcohol for cleaning the prints. Some people use Mean Green, or Simple Green cleaners instead of harsh chemicals.

$ An extra curing station (if you don't use sunlight) with a UV light inside. We have a 5 gallon bucket lined with aluminum foil, and a UV light pointing down through a hole cut in the lid. Inside the bucket we have a cylindrical container filled with water, to cure the prints underwater. This stops the resin from reacting with the air, and makes our prints come out nice and smooth, with no residual tackiness.

Figure 9-8 shows a simple 5 gallon bucket with a 405nm UV light positioned downwards into the aluminum lined chamber. Inside of the bucket is another container filled with water where the prints are post-cured for anywhere from 2 minutes to 20 minutes depending on the resin's particular needs

MATERIALS AND STORAGE:

- ($) An airtight bin you can store your filament in to control moisture absorption. Five-gallon paint buckets (with lids) are a great option here!

- ($) Rechargeable desiccant canisters to keep the air in the bin as dry as possible.

- ($) A suitable supply of "feedstock." This would be filament for the FDM printers or resin for the SLA printers. For FDM, a starting purchase of just one spool will last quite a long time, and somewhere around 1 liter of resin will be a good starting amount.

- ($) Large storage unit (preferably with wheels) to store materials such as unused filament, tools, cables, acetone, etc.

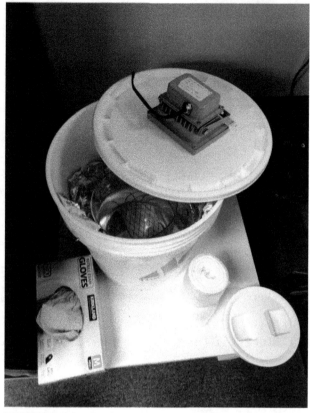

FIGURE 9-8: A 5 gallon bucket with a UV light on top, and a water-filled container on the inside, surrounded by aluminum foil.

MISCELLANEOUS:

- ($) Good lighting (via light fixtures or natural, but indirect, light). Remember that direct light from a window can negatively affect the ambient room temperature.

- ($) Indoor fans/heater (if the room doesn't have a way of regulating the temperature already).

- ($) Zip ties for securing wiring when repairing or tinkering with the printer parts. (But hey, you are a maker and makers get creative with the tools they need. If you have string or wire tie-backs... use them.)

- ($) Good WiFi access in your makerspace, so you can easily access online tutorials and help on your laptop, desktop computer, tablet, or smartphone, as needed.

We realize that this is a long list of "things you should consider having," but you probably own many of the items on this list already, and many of the things you probably don't have are commonly available (see

FIGURE 9-9: Common tools and hardware you might already own. Image by Moreno Soppelsa

Figure 9-9). In the beginning, get only what you absolutely need and purchase new supplies and equipment as you go. It may seem like a lot of prep work, but creating a space in your home dedicated to creativity and exploration can help provide a sense of purpose and fulfillment and is worth the investment.

Now that your space is set up (or on its way there), you will need to create some models to feed to your new 3D printer. The following chapters are focused on some of the free tools you can use to create not only your first models, but incredibly advanced consumer prototypes...your choice!

PART III: DIY TUTORIALS & CAD TROUBLESHOOTING

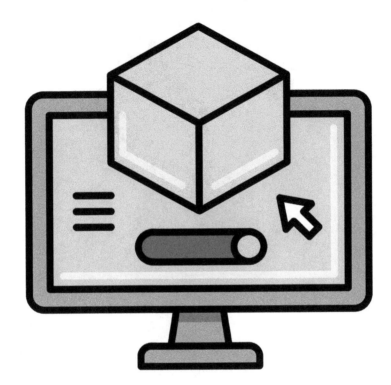

CHAPTER **10**

GETTING STARTED WITH TINKERCAD

FIGURE 10-1: A yummy-looking ice-cream cone modeled by a 4 year-old in Tinkercad

In the next three chapters we will take you step-by-step through the basics of creating a 3D model from scratch using three CAD modeling software programs: Tinkercad, Meshmixer, and Fusion 360. You can view them respectively as beginning, intermediate, and advanced programs, though we have taught students as young as nine years old to use Fusion 360.

These CAD programs have the benefit of being free (or near free), so trying them out is a great way to start. These beginning tutorials are short enough that you should be able to complete each one in an hour, but feel free to move through them at your own pace.

Three-Button Mouse Recommended

All CAD programs are optimized for a three-button mouse with a scroll wheel. When you approach CAD modeling you can use a trackpad on a laptop, but the experience is less than optimal.

Tinkercad is an online CAD modeling platform from Autodesk that is free for anyone to use. It's known as a quick and easy way to create 3D models and is simple enough for beginners and kids to pick up easily. In fact, in our educational courses, we always start beginning students, regardless of age, on Tinkercad. See **Figure 10-1** for an example of the type of simplistic model you can create in Tinkercad.

The platform is entirely browser based (so there's no software to install), which allows for your work to be instantly saved to the cloud and available on any other computer. Tinkercad runs best in modern browsers such as Firefox, Safari, or Chrome. Keep in mind that slow Internet speeds can negatively affect the overall experience.

One great thing is that Tinkercad is aimed at children and new-to-design students in general. Tinkercad is COPPA (Children's Online Privacy Protection Act, US law) compliant, meaning that there are a couple of extra steps when setting up a child's account on Tinkercad, but that child's information will be kept safe and treated appropriately by Autodesk. **Figure 10-1 on the previous page** shows an example of what students can create.

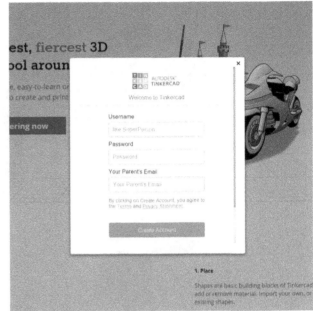

FIGURE 10-2: Tinkercad account sign-up requiring parental permission for users under 13 years of age

SETTING UP AN ACCOUNT

In order to access Tinkercad you will need to have an Autodesk account. The good news is that this program is free and the promotional emails you might receive from Autodesk are minimal.

If a child under 13 years of age is signing up to use Tinkercad, their user ID request will get routed into a COPPA (Children's Online Privacy Protection Act) compliant subsystem inside of Autodesk that requires an adult to verify that the child is allowed to create an account **(Figure 10-2)**.

The child's account will be in a special part of the Autodesk infrastructure where it will comply with the COPPA requirements. There will be restrictions on sending marketing emails as well as restrictions on sharing that child's account information with third parties. This is a good thing for the protection of children, but keep in mind this extra step can add up to 24 hours of waiting time for kids to join the Tinkercad community. This is especially important to remember if you are using Tinkercad in an educational setting where parents will need to pre-register their kids on the platform.

BEFORE YOU PICK UP THE MOUSE

We recommend you first use pencil and paper to draw your idea out. The basic premise for all CAD design is understanding that shapes interact with other shapes to create new shapes. This concept sounds easy enough, but if you are just starting to learn CAD, the best place to start is having a solid understanding of what you are trying to create. For most people, drawing it out on paper is more intuitive. Don't worry if it's not to scale or if your drawing of a house looks more like a shoebox. You will benefit greatly by having this visual guide as a reference.

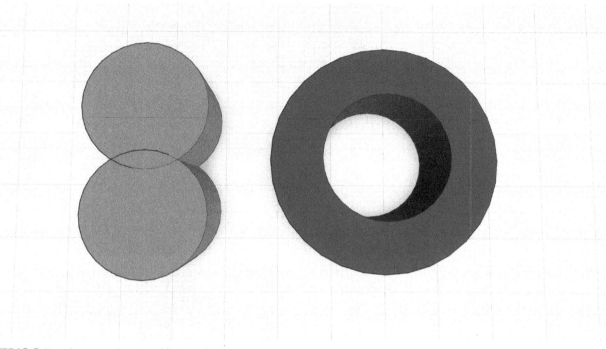

FIGURE 10-3: Line drawings showing additive and subtractive properties

"Boolean" Defined

The important CAD term "Boolean" refers to the addition or subtraction of shapes from other shapes to accomplish an end goal. Tinkercad is adept at performing Boolean unions (adding) and Boolean differences (subtraction), which are the core of how to start modeling your design. In the following tutorial, we'll use Tinkercad to model a simple nameplate. This tutorial will go over three key procedures: adding shapes on the "Workplane," turning shapes into "holes" or "cutting tools," and downloading your model for 3D printing.

If you draw a circle on a piece of paper, and then you draw another overlapping circle just touching the side of that one, it looks like a figure "8." This is an example of "additive" construction—adding two things together to create something new. Conversely, if you draw a circle and then draw a smaller circle inside it, and "remove" the inner circle, you end up with the letter "O." This is subtractive construction, i.e., building something new by taking away material. Both examples are shown in **Figure 10-3**.

CREATING SHAPES IN TINKERCAD

When you first start Tinkercad we highly recommend you go through the software's free set of online tutorials. These great lessons will walk you through Tinkercad's features at a pace you can set for yourself. They are available at the top of your Tinkercad account page under the "Learn" link. You can always click the Tinkercad logo shown in **Figure**

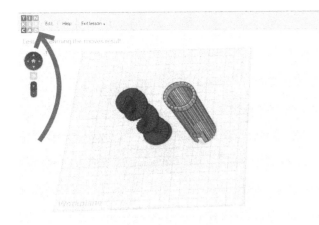

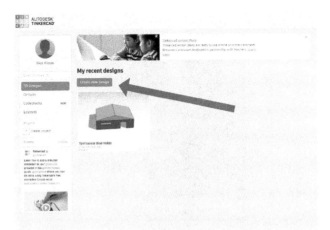

FIGURE 10-4: The Tinkercad logo takes you back to your home page

FIGURE 10-5: The "Create new design" button lets you begin your new project

10-4 located at the top left to return back to your main account page.

When you return to your home page, click the "Create new design" button **(Figure 10-5)** to start a new design.

You are presented with a blank canvas **(Figure 10-6)**, upon which you will create a masterpiece! Or, as is true with most model design, you will create something that you will call "your first try" and then figure out how to make it better over time.

FIGURE 10-6: The blank canvas that is called the Workplane

Next to the Workplane, on the right hand side of the screen there is a menu bar filled with premade shapes you can use. These shapes will help you create your 3D model.

Click and hold your mouse button down on the red "box" in the menu, and drag it out onto the Workplane as shown in **Figure 10-7**.

You now have a basic shape you can adjust, and it will be automatically selected. If you click your red box and hover your mouse over the small white

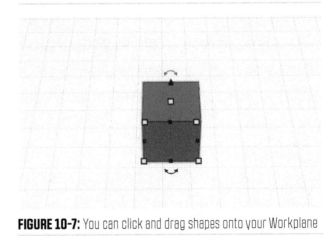

FIGURE 10-7: You can click and drag shapes onto your Workplane

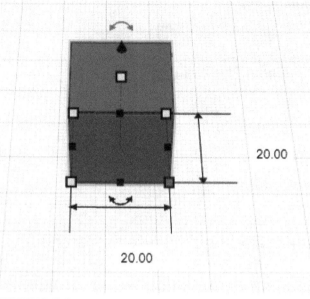

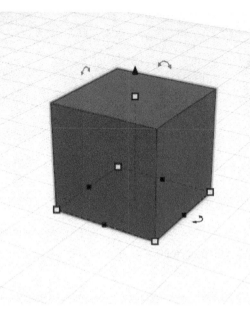

FIGURE 10-8: Dimensions pop up when you hover your mouse over an object's corner.

FIGURE 10-9: Various handles on an object in Tinkercad

squares at the corners, dimensions will pop up as shown in **Figure 10-8**.

After clicking the box, small white squares appear at the corners, allowing for resizing. The default dimensions of the box are 20mm wide by 20mm deep by 20mm tall. Clicking and dragging those small squares will resize the cube in that particular direction. Carefully roll your mouse over a tiny white square, and it will turn red to indicate that you can click-hold it. Sometimes it is hard to really click that small white square, so have patience!

These tiny white boxes and other shape modifiers are called "handles," and they allow you to resize and reshape any object in Tinkercad. See **Figure 10-9**.

HERE ARE DESCRIPTIONS OF THE VARIOUS HANDLES ON YOUR CUBE:

- Four white boxes on the bottom corners to resize the shape in the XY plane
- Four black dots on the midlines between the white handles that allow you to resize the shape by the edges instead of the corners, locked in to the direction that face is facing
- A white box on top of the object to make the shape taller/shorter in the Z direction
- A black arrow on top of the object to physically raise the object off of the Workplane it is sitting on
- Three rotation arrows that allow you to rotate the shape in any of the three directions (also note that you might not be able to see one of the rotation handles until you rotate your view around to the side of the box by holding down the right mouse button and dragging)

ROTATING AN OBJECT

There are two ways to rotate your object: by using your mouse or by using the application itself. If you look in the top left-hand corner of the main Tinkercad screen, you will see a cube and some buttons as shown in **Figure 10-10**.

Clicking on any face in the view cube will orient to that face. Click-dragging on the cube will rotate your view. Clicking the plus and minus (+ and –) signs allows you to zoom in and out. The 3D "box" symbol below the minus button switches from perspective or orthographic view. Perspective view makes parts of the model appear larger when they are facing you. Orthographic view keeps all faces locked to their actual size, no matter how close they are to you (no distortion).

The mouse is far easier to use than any keyboard shortcut. Here is a quick guide for the mouse control scheme, as shown in **Figure 10-11**.

Holding down the right mouse button will allow you to rotate the scene to see your design from different directions. Zooming is accomplished by the mouse wheel. Holding down the middle mouse button and dragging will pan the scene.

CHANGING THE SHAPE

Now that you have the basic movement down, it is time to start creating your idea! If you resized the original red box by clicking and dragging on the handles, you will have noticed that the measurements have changed. These measurements can be directly sent to a 3D printer and will be 3D printed accurately!

FIGURE 10-10: The navigation panel in Tinkercad

RESIZING THE RED BOX

Resize the box to dimensions of 5mm high by 40mm wide by 60mm long, as shown in **Figure 10-12** on the next page.

You now have a flat rectangular base. We'll use this as an example of how to use shapes "negatively" as cutting tools to create new shapes. This is a key concept not just in Tinkercad but in all CAD programs.

In the right-hand menu bar list of tools, scroll down to the "Symbols" toolbox and drag a heart shape onto the Workplane. Anywhere is okay for now.

Click the heart shape and make sure it's selected before you perform these steps:

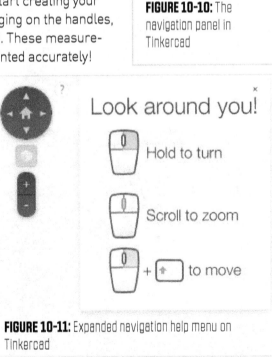

FIGURE 10-11: Expanded navigation help menu on Tinkercad

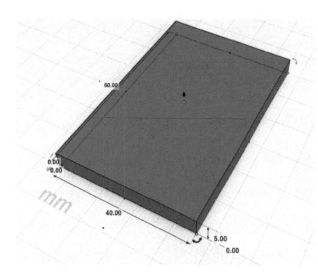

FIGURE 10-12: Resized original red box using handles

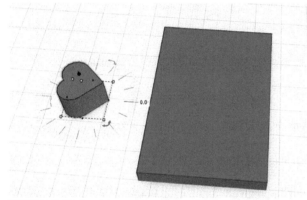

FIGURE 10-13: Stock heart shape rotated to a specific angle

Helpful Shortcuts

If you want a keyboard shortcut to zoom into an object, just click the object and hit the "f" key to "focus" and the object will fill your screen. If you want to get back to the default view, you can click the "home" icon (in the disc) and your view will be reset. In general you can navigate more easily with your mouse than your keyboard.

Using the tiny white box handles, make it 10mm high. Use the rotation arrow in the navigation tool to rotate it 90 degrees counterclockwise. Your Work plane should look like **Figure 10-13**.

As you click and drag on the rotate handles a grid will appear to help you find the right degree-angle. As you are holding down the left mouse button and rotating the object (if your mouse is outside of the rotation circle) you will move in single-degree increments. If your mouse is on the inside of the rotation circle, your object will jump in 22.5-degree increments. Or in other words, by these increments: 22.5, 45, 67.5, 90, etc.

You can rotate your view so that the heart is right-side up. Drag your heart inside the rectangle near the corner, as shown in **Figure 10-14**.

Now comes the fun part! Click the heart to select it. In the top right of your screen look at the window entitled "Shape," as shown in **Figure 10-15**.

Clicking the left box will allow you to change the color of the pieces you selected, but for this step, click the "hole" option (shaded gray stripes) and see what happens to the heart!

The heart has now become gray striped to denote that it will be used as a cutting tool. Your heart will look like the one shown in **Figure 10-16**.
You are one step away from creating your first subtractive shape!

GROUPING AND UNGROUPING SHAPES

Tinkercad is a semi "parametric" tool, which means that behind the scenes, everything you are interacting with on the screen is actually based on

FIGURE 10-14: Positioning the heart in the upper-left corner

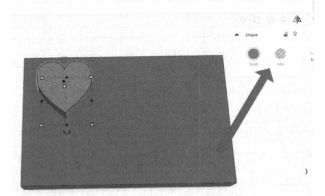

FIGURE 10-15: Shape window when an object is selected

FIGURE 10-16: Heart changed to "hole" to denote it is now a cutting tool.

simple mathematics. We say "semi" because Tinkercad remembers some aspects of your environment (like the grouping of objects as we will discuss below). It is not a full parametric modeling system where every change you make is recorded in a timeline. Fusion 360 (discussed in Chapter 12) is a full parametric modeler, in case you want more capability / complexity. The "parameters" that you set for length, width, and height determine the shape of an object. Also, Tinkercad has the concept of "nesting" or grouping objects together. Objects that are grouped together have special abilities that non-grouped objects do not. Grouping objects in Tinkercad saves you a lot of time in that you can move all of the grouped objects together all at once. And here is another great feature of the grouping: you can create a cutting tool (like the heart), then group the objects together, next you will actually see the effect of the cutting tool in your model on all grouped parts!

Let's try the grouping function. In the upper-left corner of your Workplane, click and hold your left mouse button and drag a box around both the heart and the rectangle, selecting both of them. Alternatively, you can click on the box, and then hold the "Shift" button down and click on the heart to select both objects.

In the top right of your screen, click the "Group" button as shown in **Figure 10-17**. The group button looks like a square that has been merged with a circle.

You will see a result that looks like **Figure 10-18**.

The heart is now cut out of the rectangle! With Tinkercad being a partial parametric modeling program, this also means that the operation you just completed of cutting the heart out of the block is "non-destructive." That means that you can go back in to change things around. This will not disrupt other objects in your design. It's like it has a memory for what you did in layers.

If you want to change the group you just created, double-click with your left mouse button on the rectangle and "enter". You can change the placement of the heart, or really any component that is in that group. When you click outside of the grouped shapes, the various cuts are performed based on the new locations of the shapes. Conversely, you can also click the "Ungroup" button next to the right of the Group button to ungroup selected shapes.

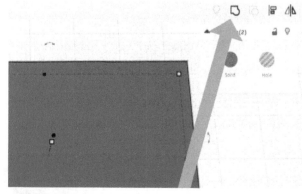

FIGURE 10-17: Location of the Group button

FIGURE 10-18: Result of grouping heart cutting tool and rectangle

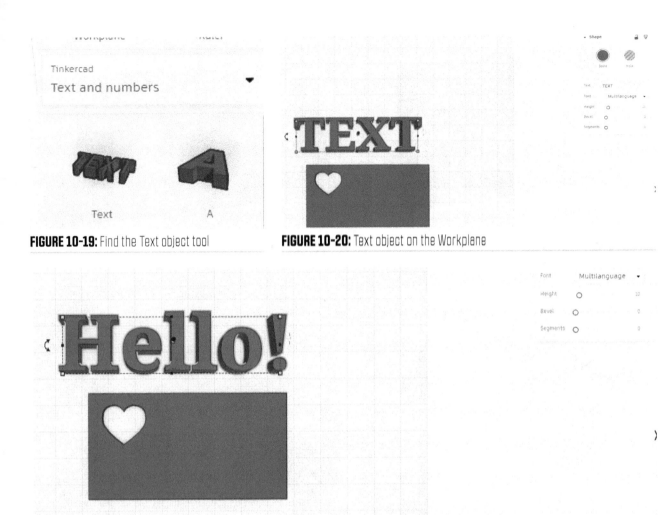

FIGURE 10-19: Find the Text object tool

FIGURE 10-20: Text object on the Workplane

FIGURE 10-21: Text changed to "Hello!"

FINISHING YOUR NEW CREATION IN TINKERCAD

The next steps of this tutorial will demonstrate how to further customize your heart-plate using other customization tools:

On the right-hand panel, select "Tinkercad → Text and Numbers" and select the "Text" tool as shown in **Figure 10-19**.

Drag the Text object out onto the Workplane as shown in **Figure 10-20**.

Use the Inspector tool in the top right to change the "Text" to something else as shown in **Figure 10-21**. Drag the text into your nameplate (the rectangle) and resize if necessary as shown in **Figure 10-22**.

FIGURE 10-22: Resized "Hello!" placed on the nameplate Resized "Hello!" placed on the nameplate

FIGURE 10-23: "Hello!" turned into a "hole" cutting tool

FIGURE 10-24: End result of cutting after objects are grouped

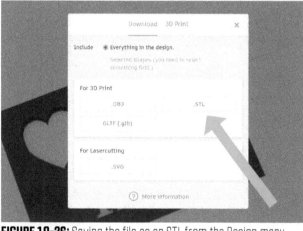

FIGURE 10-25: Download menu when clicking on "Export"

FIGURE 10-26: Saving the file as an STL from the Design menu

With the text "Hello!" still selected, use the Shape window in the top right to turn the Text object into a "hole" as shown in **Figure 10-23**.

Using your mouse, drag a box around all of the objects. Next, in the top right of the Tinkercad window, click the Group button at the top right of the screen to see the result shown in **Figure 10-24**.

Click "Export" in the top right of the Tinkercad window as shown in **Figure 10-25**.

Click "For 3D Print" and then choose ".STL" as

shown in **Figure 10-26**. This is the file format you could 3D print!

Congratulations! When you have an STL file from Tinkercad, you can load that file into your 3D printer or upload it to an online service bureau where they will 3D print your design for you.

Now that you understand the basics, we recommend that you spend time playing around with Tinkercad on your own until you feel comfortable with the features and controls. You will learn best by practicing, so don't be afraid to experiment.

Tinkercad is a great program for young and old alike. The tutorials inside of Tinkercad help learners understand how to use objects not only in additive ways, but also in negative space as well. There is nothing better than seeing someone new to CAD modeling measure out some shapes on a sheet of paper, then CAD modeling them in Tinkercad, and finally 3D printing the files. **Figure 10-27** shows a proud new CAD modeler holding some of his creations...way to go!

WHAT'S NEXT?

In Chapter 11 we give you another CAD tutorial. We'll

FIGURE 10-27: A proud beginner CAD modeler holding physical printouts from his Tinkercad creations. A rocket, star, arrowhead and more!

FIGURE 10-28: The STL file loaded into Meshmixer

FIGURE 10-29: A free-form model in Meshmixer

look at Meshmixer, another fun and free CAD modeling software program from Autodesk. This one has a medium level of complexity but allows you to do far more with the software than what you can do with Tinkercad.

If you already have Meshmixer installed on your computer, you can double-click (or import) the Tinker-cad-generated STL file you just created into Meshmixer for further editing, as shown in **Figure 10-28**.

Some changes are needed to make the file more easily editable in Meshmixer. You can do this by "remeshing" (using the Meshmixer commands Ctrl-A to select the model, then Edit → Remesh → Linear Subdivision a couple of times) to make the model more complex so that Meshmixer has enough triangles to work with. But, once the model has enough triangles in it, you can do some fun things in Meshmixer **(Figure 10-29)**.

When you're ready for the Meshmixer tutorial, just turn the page!

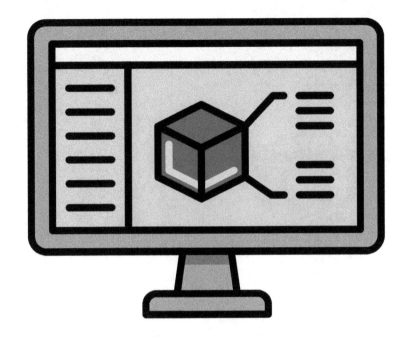

11

GETTING STARTED WITH MESHMIXER

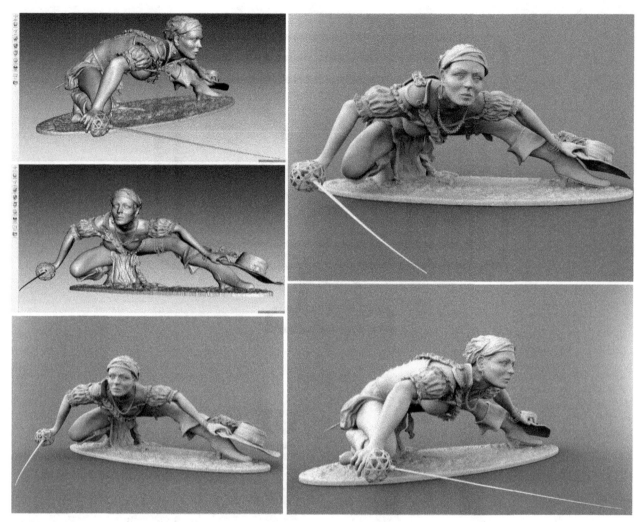

FIGURE 11-1: The two gray figures (top left) are created in Meshmixer; the other three views have had color and shading effects added in an external program to make the 3D model look nicer in image form (Artwork entitled "Madion" by Gunter Weber/"MagWeb" on the Meshmixer forums)

Meshmixer is another free program, and it is amazingly powerful. It's so useful that, if you 3D print a lot, you may find yourself using Meshmixer before every print. Don't be fooled by the free price and by our "medium-level" skill rating; it can be just the right toolbox beginners need to print and create models, as well as something advanced users can use to create digital masterpieces. Check out the model in **Figure 11-1** for an example of the kind of sculpted 3D model you can create with Meshmixer.

Here are a few reasons why Meshmixer is one of our most popular online 3D printing and 3D modeling courses on **LinkedIn Learning /Lynda.com**:

⊘ Meshmixer can be used not only to create all kinds of 3D models, but also to help you translate those models into physical products. This ability is unmatched by any other software program.

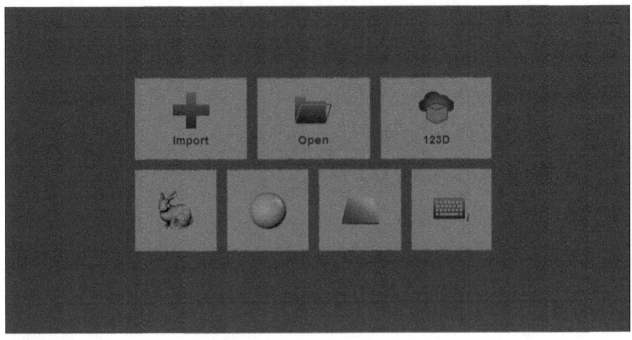

FIGURE 11-2: Meshmixer Start menu

- Even if you never end up 3D printing what you create, Meshmixer is a powerful standalone digital creation tool.
- Meshmixer helps you get the most out of your own 3D printer at home.
- Students, professionals, artists, hobbyists and designers, and beginners and advanced modelers can all benefit from learning Meshmixer.
- Meshmixer can be taught to ages 9 and up.
- There are over 80 tools in the program, many of which are easily applied to your model.

GETTING STARTED

Meshmixer is not browser-based like Tinkercad, so you will need to download the program by visiting the official website. Go to **http://www.meshmixer.com** and follow the instructions to download and install Meshmixer on your computer.

When you first start Meshmixer you will be presented with a menu of starting choices, as shown in **Figure 11-2**.

BEFORE WE GET ANY FURTHER, LET'S GO OVER SOME BASIC ASPECTS OF MESHMIXER:

- Left-clicking your mouse will select things on the screen, or act as the brush when sculpting
- Holding down the right mouse button and dragging your mouse will rotate the object
- Holding down the middle mouse button will pan the object

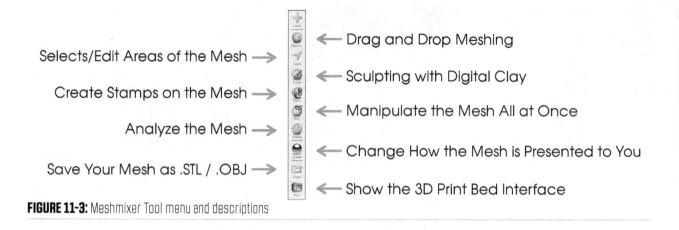

Selects/Edit Areas of the Mesh →

Create Stamps on the Mesh →

Analyze the Mesh →

Save Your Mesh as .STL / .OBJ →

← Drag and Drop Meshing

← Sculpting with Digital Clay

← Manipulate the Mesh All at Once

← Change How the Mesh is Presented to You

← Show the 3D Print Bed Interface

FIGURE 11-3: Meshmixer Tool menu and descriptions

Figure 11-3 shows the left hand side of the Meshmixer screen with labels of the tool functions. You are able to access all the tools of Meshmixer from these main icons.

STARTING THE TUTORIAL

From the Start menu, click the rabbit in the lower-left corner. This will load up the iconic "Stanford Bunny," which was 3D scanned from a small statue in 1993 at Stanford University. Once you click the icon, a bunny will appear on the screen. You now have a 3D model to work with.

Now that you have the bunny on the screen, though, you will need to "fix" it. Because this is a 3D model that was created by 3D scanning a real-world object, the scanner was only able to capture the exposed surface areas of the bunny, and not the parts where it was sitting on the table. To 3D print this bunny, this "missing and open" part of the scan will need to be closed. 3D printing requires a closed, airtight model. Fortunately, Meshmixer has some powerful tools that will evaluate and "fix" 3D models to make them more 3D printable.

BEGIN BY CLICKING ANALYSIS → INSPECTOR.

You will see a blue "pin" popping out of the bottom of the bunny as shown in **Figure 11-4**. This indicates that there is an issue with the mesh in that location. You can easily see the big hole in this 3D model, but some holes are harder to see, making this tool very valuable. Meshmixer will find errors such as holes and small disconnected parts with the Inspector tool. Make sure to rotate your model all the way around with your right mouse button to see all pins that might be hidden from view.

Next, click the blue pin (or the "Auto Repair All" button in the Inspector tool), and you will see the bottom of the bunny fill in.

Click Done, and your bunny will be corrected for 3D printing as shown in **Figure 11-5**.

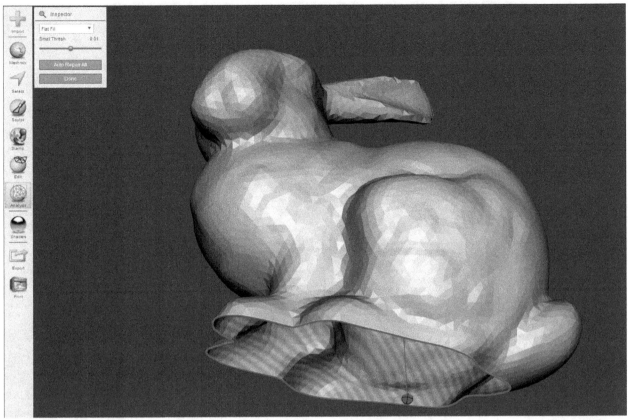

FIGURE 11-4: Analysis and Inspector tools on the bunny model

SCULPT WITH DIGITAL CLAY

As we mentioned, Meshmixer has more than 80 amazing tools. One of the most fun to play with is the Sculpt tool. Let's make the bunny more interesting by creating some wings to allow it to fly:

CLICK THE SCULPT TOOL ON THE LEFT MENU. Click Brushes and select the **Draw** tool. The **Draw** tool allows you to add to and create volume in the model, as if you were applying more "clay" to it.

Before you start modifying the bunny, notice that there are settings that affect your brushes on the left-hand side. These options have a dramatic effect on how the tools work. For now, open up the settings windows to access the properties and

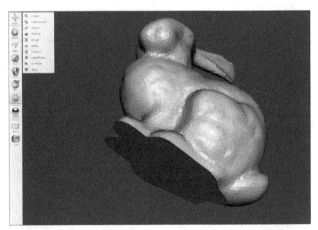

FIGURE 11-5: Bunny model is fixed using the Inspector tool

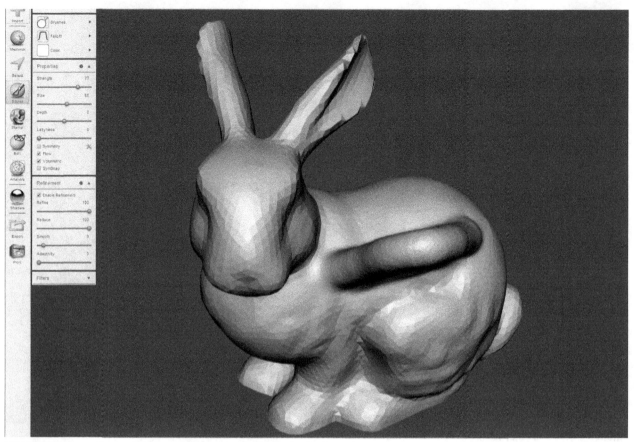

FIGURE 11-6: Properties and refinement areas are adjusted in the Sculpt tool

refinement areas **Figure 11-6**, and adjust the settings to match the following.

IN THE PROPERTIES AREA:

Strength: 77
Size: 55
Depth: 0
Flow and Volumetric should be checked

IN THE REFINEMENT AREA:

Enable Refinement should be checked
Refine: 100
Reduce: 100
Smooth: 9

FIGURE 11-7: Both wings have been added to the bunny

Now, rotate your view so that you are looking at the left shoulder of the bunny. Hold your left mouse button down on the shoulder area, and while keeping the left mouse button down, start moving back and forth in a line as in #fig10_6. If you spend a bit more time in the center than you do at the edges, you will be slowly building up a wing on the bunny.

Once you have drawn one wing, rotate your view to the other shoulder with the right mouse button, and "draw" another wing so that your bunny can properly fly as shown in **Figure 11-7**.

ADDING SUPPORT STRUCTURES

We talked about the need for support structures in

A Note on Sculpting

Many sculpting commands (including the Draw command) will build toward you, in other words toward the viewpoint from which you are looking at the model. So, as you are "drawing" the bunny's wing, it might seem as if nothing is happening... but if you rotate around, you will see that the wing is being "drawn" toward you.

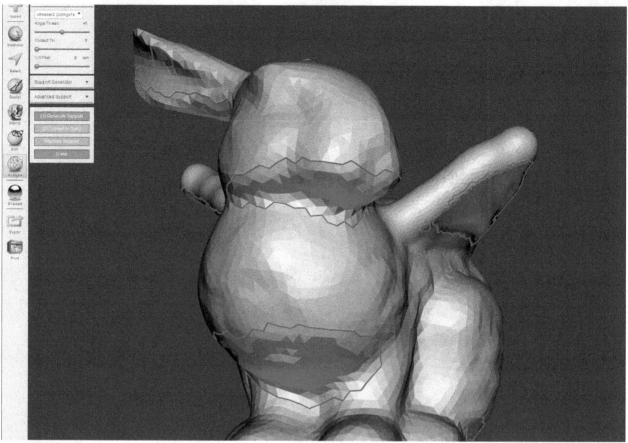

FIGURE 11-8: Adding support structures to the bunny

previous chapters, and Meshmixer is a truly unique program in its ability to help you successfully 3D print. One standout feature of Meshmixer is the creation of support structures that will support overhanging areas of your model so that they will not droop during 3D printing. Let's create some of those support structures now:

ON THE LEFT-HAND MENU, CLICK ANALYSIS → OVERHANGS.

Change the top option from Custom Settings to Ultimaker2. You will see that certain places on your model have turned red, as shown in **Figure 11-8**. Those are the areas that qualify for the generation of support structures, based on the settings you entered in the menu on the left. You can change these settings to what works best once you learn the capabilities of your specific 3D printer. (The Ultimaker2 printer was chosen for this example, but a different choice may be appropriate for your specific printer. You can always use one of the prefigured settings as a starting point for your own printer's needs.)

CLICK THE "[1] GENERATE SUPPORT" BUTTON.

You just told the program to create support structures to keep your model in place during the printing process.

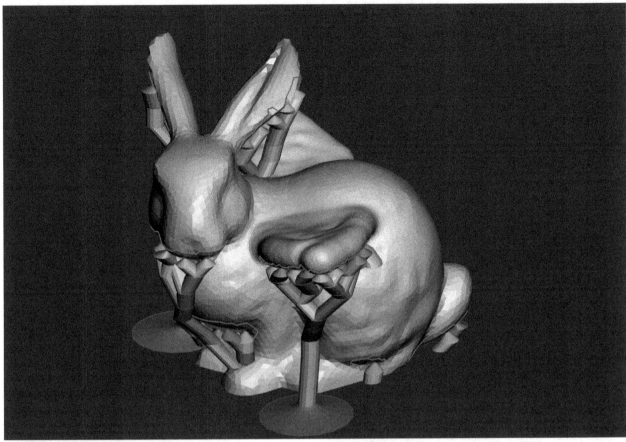

FIGURE 11-9: Support structures have been added to the bunny for the 3D printing process

Meshmixer will think about the support structure generation for a few seconds, and then support structures will appear as shown in **Figure 11-9**!

As we mentioned before, these support structures can be edited and removed at will, or regenerated after you change your settings. For you, learning how well your 3D printer prints at specific angles and with specific materials will be part of the learning curve.

CLICK "[2] CONVERT TO SOLID" TO LOCK IN THE SUPPORT STRUCTURES AS ACTUAL 3D MODELS THEMSELVES.

A menu will pop up, asking if you want to create a "New Object" or "Replace Existing" object on the

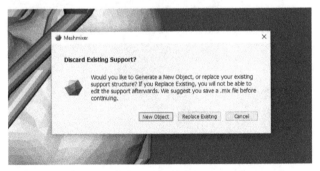

FIGURE 11-10: Replacing the existing object using the pop-up menu

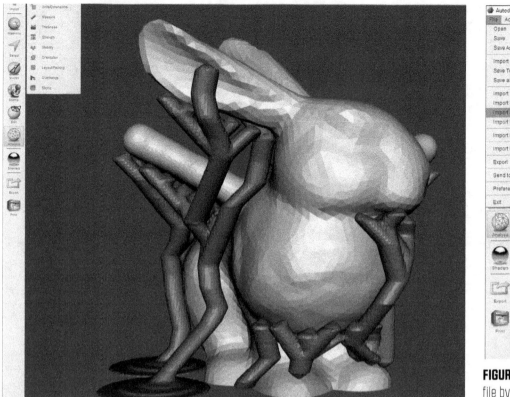

FIGURE 11-11: Creating new support structures by changing the settings

FIGURE 11-12: Creating a new file by choosing Import Sphere

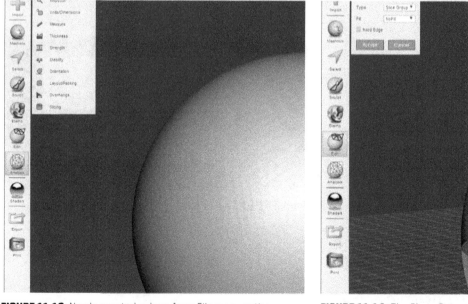

FIGURE 11-13: Newly created sphere from File menu option

FIGURE 11-14: The Plane Cut pop-up window from the Edit menu option

screen. Click Replace Existing as shown in **Figure 11-10**.

You will be brought out of the support structure generation utility and back into the main interface. You can click the **Export** button on the bottom left of the side menu to save this model for 3D printing if you desire, or you can play around with editing the bunny or even the new support structures as shown in **Figure 11-11**.

CREATING A SPHERE

We've shown you how to fix a model, use the Sculpt tool, and generate supports. Now let's change gears for a moment and practice on more Meshmixer tools. Let's leave the bunny behind by clicking the top-left File menu, and then selecting **Import Sphere**, as shown in **Figure 11-12**.

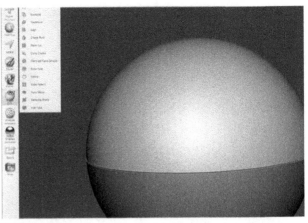

FIGURE 11-15: Using the Plane Cut option on a sphere to create FaceGroups

Meshmixer will ask you if you want to "append" or "replace" the bunny. For this step choose "replace." You will now have a sphere, as shown in **Figure 11-13**.

You will be creating some very cool jewelry in this exercise. But we first need to start with some base geometry:

Start by segmenting the sphere a few times. (You'll discover why later on.) Click the Edit button on the left-hand menu and you will see a Plane Cut pop-up window, as shown in **Figure 11-14**.

SELECT THE OPTIONS IN THE POP-UP BOX TO:

- Slice Groups

- NoFill

Click **Accept**.

You will see a ball that is gray on the top side and colored on the other, as shown in **Figure 11-14**. (If you don't see multiple colors, hold the space bar down, and select the other option under the "Colors" menu that pops up.) The "Slice Groups" command splits the sphere into two separate "FaceGroups." FaceGroups are simply areas of the mesh that allow you to manipulate them without touching the other parts of the mesh. They are denoted by a different mesh color. Your color might be different than the one shown in **Figure 11-15**.

Click the **Edit** tool again and then **Plane Cut**.

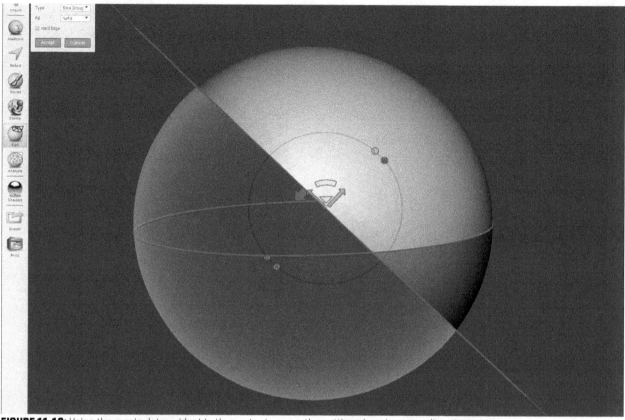

FIGURE 11-16: Using the manipulator widget in the center to move the cutting plane in a new direction

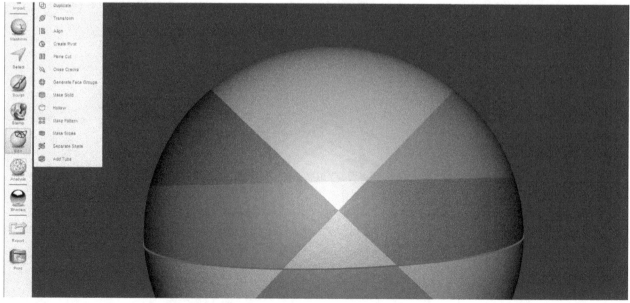

FIGURE 11-17: Sphere with many FaceGroups

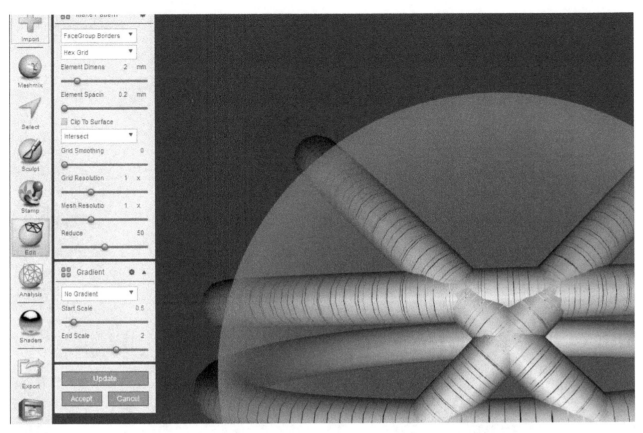

FIGURE 11-18: FaceGroup Borders option on the sphere

Choose the same options as before: Slice Groups and NoFill.

You did this before, but this time either click and drag a line with your left mouse button in another direction, or click the various arrows and bars on the "manipulator widget" in the center to move the cutting plane to a new direction as shown in **Figure 11-16**.
Click **Accept** and you will see a new FaceGroup appear.

Do this procedure two or three more times, and you will get a sphere that looks like the one shown in **Figure 11-17**.

Here is the fun part! "Make Pattern" is one of the flashiest and most powerful tools in Meshmixer. It is what you will use to make a piece of jewelry out of this colorful sphere.

Click the **Edit** button and go to Make Pattern.

Switch the drop-down menu on the left hand side to FaceGroup Borders (very top).

FIGURE 11-19: Borders of FaceGroups made into a pendant

You will see a preview of what will happen to your mesh, as shown in **Figure 11-18**.
Click **Accept**.

You now have a mesh that follows the borders of those FaceGroups and can be turned into a complex pendant, as shown in **Figure 11-19**. You could have this piece 3D printed in plastics or metals by online service bureaus. This would be considered a more advanced model to print at home, due to the amount of support material you would need, but you certainly could print it yourself.

This tutorial has been a brief introduction to the power and versatility of Meshmixer. We invite you to play around with the other tools in this program to see what you can create. To further your understanding, be sure to also check out the online tutorials provided by Meshmixer on the Autodesk 123D YouTube channel. On the education page of our website we have an award winning 16 hour online course on Meshmixer as well: **www.HoneyPoint3D.com/education**

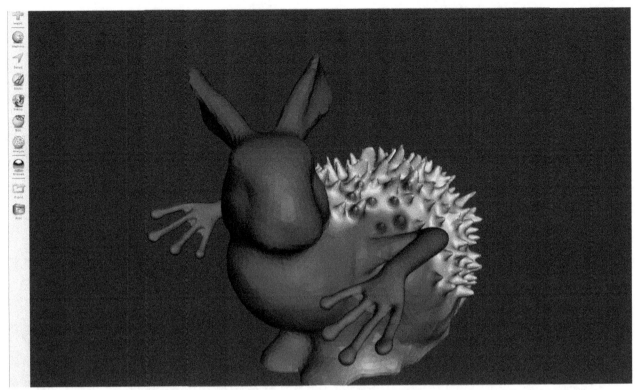

FIGURE 11-20: Bunny shape with new arms, added colors with "Paint Vertex" tool, and Sculpt → Surface → Draw++ Stencil applied

Online Tutorials

If you are looking for more in-depth online training we recommend our course called "3D Printing and 3D Design Using Autodesk's Meshmixer" **https://www.honeypoint3d.com/education/**. There are 159 lectures and over 16 hours of step-by-step instruction you can pace at your own level. In this course you will learn some other very popular, fun, and easy tools like "Dragging and Dropping," which is exactly like it sounds. You can pick pre-created shapes/objects and drop them onto your starter model as shown in **Figure 11-20**.

In the next chapter, we will introduce you to Autodesk Fusion 360, which is considered a professional-level CAD program. It is very powerful software but is nonetheless very approachable for individuals.

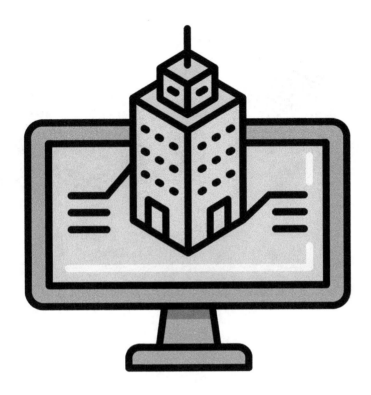

CHAPTER

12

GETTING STARTED WITH FUSION 360

FIGURE 12-1: A freely available "Open Chair" design (left), and a close up photorealistic rendering of the straps and buckles used in the chair (right). Modeled by Jasper Cohen (courtesy: https://jaspercohen.com/)

While Tinkercad and Meshmixer are good, free programs, it's pretty much guaranteed that the full fledged consumer products in your life were not designed with those tools. To create 3D models that can be sent to manufacturers for production, designers need to use more advanced tools such as Autodesk's Fusion 360. **Figure 12-1** shows an example of the complexity of objects that can be created in this program. This design allows you to download and fabricate the components of your own furniture, using 3D printed parts and commonly available components (straps, etc.).

Fusion 360 enables the creation of complicated models and costs a fraction of what other software in this league would cost. At the time of this writing, the pricing of Fusion 360 is fairly unique in the software market:

- ✅ Free for personal/student use (with some restrictions on what you can do)
- ✅ Free commercial use for startups until their business makes more than $100,000 per year and as long as you are designing products for your own company to sell. This promotion requires acceptance into the startup program, at the discretion of Autodesk sales
- ✅ After that, Fusion costs around $50/month when you prepay for a year

This fee structure allows users full access to a very powerful software program, without having to pay for an expensive commercial license that can cost thousands of dollars. In fact, most of the CAD models we create in our rapid prototyping business are made in Fusion 360. It is a powerful tool for professional work.

In this chapter, we'll provide you with a basic introduction to help you understand how the program works and how to get started using it. An in-depth tutorial covering all the rich features of Fusion 360 is beyond the scope of this introductory book. (We recommend Fusion 360 for Makers, 3rd Edition by Lydia Sloan Cline, or our online course located on our website **https://www.honeypoint3d.com/education**).

POLYGONAL VERSUS PARAMETRIC MODELING

Fusion 360 creates what are called "solid models" that have a defined volume and weight based on the material assigned to them. For example, you can make parametric models for objects that are made out of concrete and others for objects made out of rubber. Each of those models can have simulations run against them to illustrate what would happen to the objects in the real world if specific stresses were placed on them. After all, the density or other material properties in a CAD program like Fusion 360 is just another formula to calculate.

Before we jump into the tutorial, however, we want to remind you about the difference between polygonal and parametric modeling. The STL file format that 3D scanners create, and that Meshmixer uses, generates hundreds or thousands or even millions of small triangles (which are really just three-sided polygons) to create an object. These triangles create a shell of the object rather than a solid with volume. Imagine casting a fishing net over an object. The fishing net is like an STL file that shells the object.

Parametric modeling, by contrast, uses parameters to define a model. Parameters are another way of saying changeable mathematical formulae. When you look at a model on the screen you are seeing how one formula interacts with another, which is why CAD programs are able to make huge changes in the model very quickly.

Fusion 360 is a great example of a parametric modeling program. The file output of Fusion 360 can be STL (triangles) or parametric file formats (more-than-three-sided shapes) such as STEP and IGES, which are used in traditional manufacturing.

Parametric modeling allows you to go back in time and change values based on choices you may make later. These changes can be made at any time, and they will cascade throughout the design, all the while keeping the object in proper shape. For example, if you change the shape of a window on a house, the walls would adjust accordingly and you would not have to go back and make changes to them as well. That is because the models are based on parameters, and Fusion 360 recalculates all of the formulas accordingly.

GETTING STARTED WITH FUSION 360

As we mentioned, all the features of Fusion 360 are too numerous to cover in this tutorial. Being exposed to Fusion 360 is important if you want to someday advance past Tinkercad and Meshmixer. The following beginning tutorial in Fusion 360 is for anyone looking to add more advanced features to their CAD modeling software toolkit.

In the following steps you will be designing a ring. If you have not done so already, download and install Fusion 360.

You will need to create a free Autodesk account to use the software. You might have already obtained an Autodesk account if you signed up for Tinkercad or Meshmixer.

Fusion 360 is partially cloud-based. It does not strictly require a connection to the Internet for you to do the modeling, but you will have the best experience on a computer that is actively connected to the Internet. The "offline" mode should only be used when your Internet connection is temporarily not available. The main computations are done on your computer, and the only time you need the Internet connection is to save and sync your CAD design models.

All files are securely saved to the Autodesk cloud for access on any computer you might be using. That your files are stored on the cloud might be an unacceptable security risk, though, so make sure cloud storage is allowed by your own security rules. Without an active Internet connection, your experience with Fusion 360 will be much more difficult. When you start Fusion 360 for the first time, you will see the screen shown in **Figure 12-2**.

There are four different training tutorials that show the major functions of Fusion 360. We strongly advise that you go through these at some point, if not before this tutorial. You can always access these training videos by clicking the small question mark in the upper-right part of the main Fusion 360 window and then selecting Step By Step Tutorials.

If you are coming from another solid modeling program (like Dassault Systèmes: SolidWorks), Fusion 360 will give you the option to use the navigation shortcuts that are familiar to you, or use the Fusion 306 native controls. This tutorial assumes you are using the native controls.

For now, close this window and return to the main Fusion 360 interface window, as shown in **Figure 12-3**.

> **NOTE:**
>
> The images in the previous edition of this book were based on the older user interface (UI) design. These new images reflect Fusion 360's new look.
>
> These interface changes are actually beneficial! Fusion 360 is very actively improved and developed as new features and capabilities are always being added to this great software. Look for updates and improvements every 6-8 weeks (with bug fixes reported in the forums even more quickly than that).

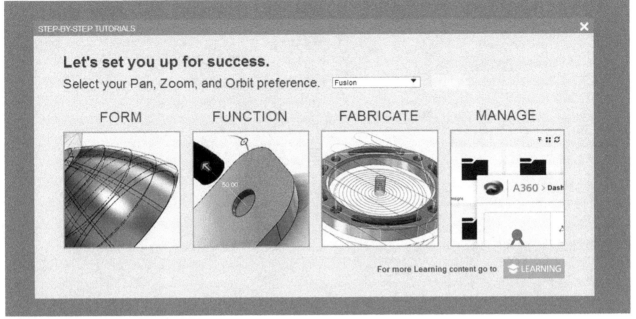

FIGURE 12-2: Fusion 360 start screen

DESCRIPTION OF INTERFACE MENU

The Fusion 360 interface is "tabbed" which means there are main categories of what you want to achieve in the left drop-down box, and then based on that selection there are common commands listed left-to-right in the top "ribbon."

Let's look at the different environments you can access via the drop-down list in the top left:

DESIGN

This is where you will create all of your 3D models. There are "tabs" for the different types of model tools you can work with. "Solid" tools are traditional shapes and sketches. "Surface" tools allow you to do advanced surface modeling, patching and stitching. "Sheet Metal" allows you to create designs that are meant to be folded into a 3D shape from a flat sheet of metal. "Assemble" allows you to create moving or fixed joints between models to create moving assemblies. "Tools" are extra options around 3D printing, measuring, exporting, and managing how you select objects on the screen.

GENERATIVE DESIGN

This environment allows you to set up conditions (like boundaries and features to be preserved) and allows Fusion 360 to create shapes and models for you! These designs are usually only manufacturable via Additive Manufacturing (3D printing), though you can state that CNC machining (subtractive process) is the desired medium. Generative design costs cloud credits to run which is an added cost Autodesk charges for some advanced operations. But don't miss checking out this highly unique feature as you can create lightweight

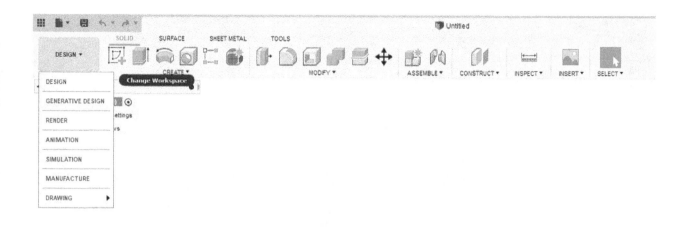

FIGURE 12-3: The main interface window of Fusion 360

parts that are the combination of many oher parts, and other cost-saving features.

RENDER
Once you have a 3D model, you can create photo-realistic images and animations based on your designs, using life-life material libraries included in Fusion 360.

ANIMATION
Allows you to create moving animations of your 3D models to showcase how they work and what they do.

SIMULATION
Once a model is created, you can actually run simulations on the computer that will help to predict failure scenarios for your objects...as long as you know the real-world forces that your objects will be subjected to. Real world forces include: the direction in which stress is applied, if any faces of your object are "locked" against other objects, static stress amount, if your object will be exposed to thermal / heat stresses, or if your object might buckle under pressure.

MANUFACTURE
Fusion 360 will create full GCode tool paths for milling machines. Once you have a 3D model you can define how you want your object milled and Fusion will create those toolpaths for you.

DRAWING
You can associate 2D drawings with your models for patent drawing purposes or to help shop-workers manufacture your product. The drawings dynamically update as you change your model.

MOVEMENT BOX

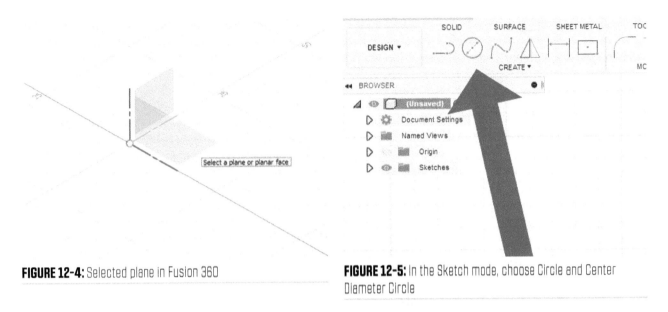

FIGURE 12-4: Selected plane in Fusion 360

FIGURE 12-5: In the Sketch mode, choose Circle and Center Diameter Circle

Also, on the top right of the screen, there is a small "view cube." This cube can be click-dragged and clicked on to rotate your model if you forget how to rotate the model using the keyboard shortcuts.

Now that you've downloaded Fusion 360 and taken a brief tour of the features, it's time to try it out.

MAKING A RING IN FUSION 360

In this tutorial you will be designing a ring that could be 3D printed either at home or through an online service bureau. The steps can be followed precisely, or you can get creative with some of the sculpting functions and add your own features. It's really up to you!

Start by creating the band of the ring:

Make sure you are in the "Design" environment, and the "Solid" tab in the top-left drop down box and in the ribbon, respectively.

The very first icon in the ribbon is a sheet of paper with a pencil and a "+" symbol. That is the "Create Sketch" tool. Click that icon. Your screen will look like **Figure 12-4**. As you move your mouse around the screen, Fusion 360 will ask you what "plane" you want to create that new sketch on.

Select the bottom plane. Now you are in Sketching mode.

In the top ribbon click the "Center Diameter Circle" as shown in **Figure 12-5**.
We are creating a jewelry ring. For the benefit of this tutorial (and because we want to make something truly custom) we measured a family member's ring finger diameter and found their finger to be 17.3mm wide.

Click on the dot that marks the center point of the modeling surface (where the red and green lines converge). As you move your mouse a circle will follow your mouse showing the diameter of the circle, as shown in **Figure 12-6**.

Before you click the button on your mouse, type 17.3 as shown in **Figure 12-7**. Then press Enter twice.

That's all you need for the circle at this point so click "Finish Sketch" in the top right of the menu bar to get back to the main modeling environment.

Now that we have a base circle, let's create some 3D geometry.

Click the small arrow in the top ribbon next to the word "Create" and select Pipe as shown in **Figure 12-8**.

Click the edge of the circle. A 3D model will be generated and a menu will pop up as shown in **Figure 12-9**.

Select the following settings in the pop-up menu:

Set both Distance settings to 1.

Change the Section Size to 1 and then click OK.

You just made a simple ring with a 1mm-thick cross-section, as shown in **Figure 12-10**.

But wait!
Let's think through the measurements of this model. Remember that you created a band to fit a specifically measured distance of a 17.3mm wide interior diameter. But we had you create a 1mm pipe on the centerline, which extends inward by .5mm

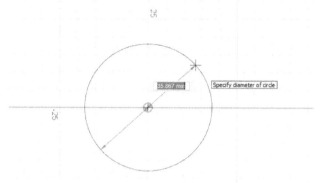

FIGURE 12-6: Drawing a circle with specific dimensions

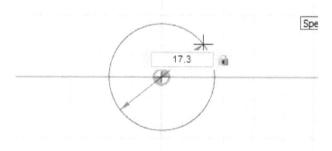

FIGURE 12-7: A sketch circle with a diameter of 17.3mm

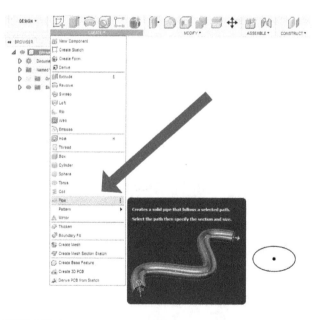

FIGURE 12-8: Selection of Pipe option under the Create icon

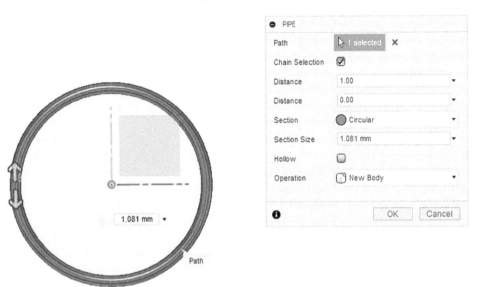

FIGURE 12-9: A 3D pipe appears with a menu option

and outward by .5mm. That essentially takes away .5mm from the inside, on both sides (removing a total of 1mm). Now the interior diameter is actually 16.3mm! If you printed this ring, it would be 1mm too small! Always conduct a logical "walkthrough" of your steps to see if they make sense.

Never fear, it is easy to fix these things in a parametric modeling program.

At the bottom of your screen you will find a "history" of all of the actions you took. Do you see the initial sketch there, as well as a pipe command? Let's go in and edit that sketch.

Double-click the first Sketch icon on the bottom left of the screen, just to the right of the Play commands, as shown in **Figure 12-11**. This will bring you back into the initial sketch you made.
To fix the model and make it larger, double-click the "17.3" number and edit it. You can also use arithmetic operations in there as well! It is valid to type either "17.3+1" or "18.3".

FIGURE 12-10: The completed band of the ring

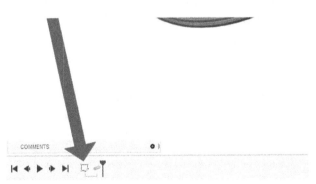

FIGURE 12-11: Arrow showing the "history" that is being built as you work in Fusion 360. Double-clicking any item will allow you to change that item.

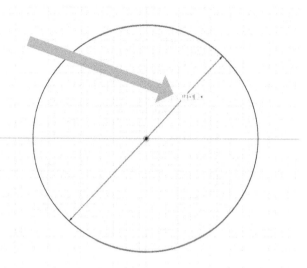

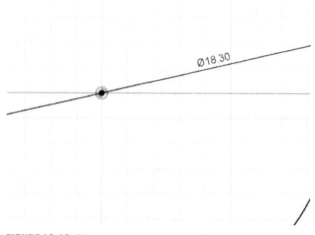

FIGURE 12-12: In the sketch environment, double-clicking the diameter number allows for editing. The field entry now reads "17.3+1."

FIGURE 12-13: Close-up view of changed diameter

Change "17.3" to "18.3" either by directly typing that number in, then press Enter, or by entering the addition formula as shown in **Figure 12-12**.

You will now be left with a circle that is 18.3mm in diameter. **Figure 12-13** shows a close-up view of the new value of the diameter.

Click Finish Sketch to return to the main environment. Notice your ring tube updated in the background to 18.3mm in diameter, but the pipe thickness is still 1mm. The effective inner size is now 17.3mm, which is what we wanted.

ADDING AN EMBELLISHMENT TO THE RING

Now, let's put a topper on that ring to make it more interesting! Click the purple grid-box in the top ribbon with a hover-over label of "Create Form" to enter into Sculpting mode as shown in **Figure 12-14**.

After entering into Sculpting/Form mode, click the top leftmost shape that looks like a 4 x 4 x 4 cube with rounded edges.

Click the same plane as you did before to create a cube on the bottom plane. In order to better see where to place the cube, find the "view cube" on the top right of the screen and click on "Top" to reorient your view to the top. Then, place the cube somewhere around the "top" of the ring, by clicking in the general vicinity of the shaded-out ring, as shown in **Figure 12-15**.

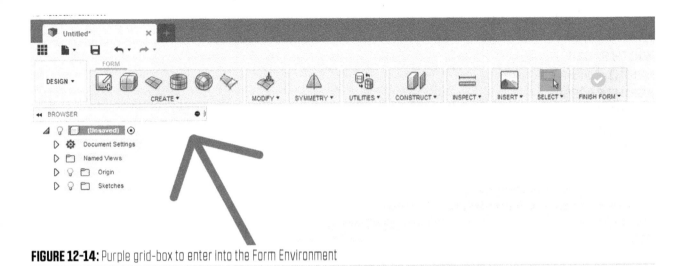

FIGURE 12-14: Purple grid-box to enter into the Form Environment

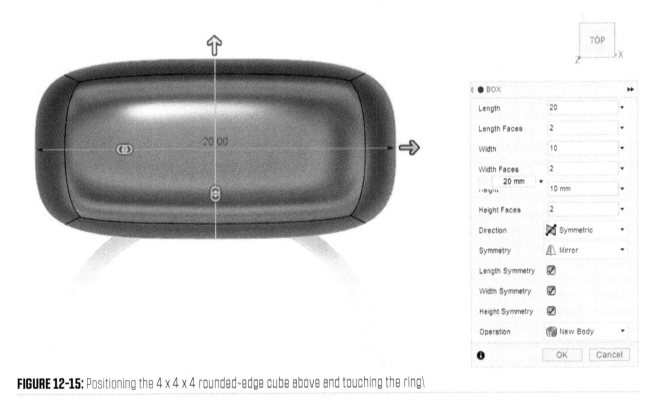

FIGURE 12-15: Positioning the 4 x 4 x 4 rounded-edge cube above and touching the ring\

Let's pretend we are jewelry designers and happen to know the settings shown in **Figure 12-15**:

LENGTH: 20mm
LENGTH FACES: 2

WIDTH: 10mm
WIDTH FACES: 2
HEIGHT: 10mm
HEIGHT FACES: 2
DIRECTION: Symmetric
SYMMETRY: Mirror
CHECK ALL THE SYMMETRY BOXES: Length, Width, Height
OPERATION: New Body

Click "OK" to accept those settings for the box.
These settings are just for this project but you can play around with changing them now, or go back into your design and change them later.

ADDING TO THE EMBELLISHMENT

You just created a sculptable box of parametric clay in the last few steps, so let's make it look unique!

Hover your mouse on top of the box. You can change your view by click-dragging on the view cube in the top right of the screen. Click one of the top edges (one of the black lines on top of the box) and it will turn blue as shown in **Figure 12-16**.

The other edges will turn yellow to indicate that symmetry is turned on; a change made to one face will affect the other edges in that object.

On the top menu bar click the Edit Form button as shown in **Figure 12-16**. A manipulator widget will appear, and as you drag those various arrows and sliders, your form will change as shown in **Figure 12-17**. You might have to use the navigation cube in the top-right part of the screen by clicking it and dragging it to rotate the view.

In **Figure 12-17** we moved the very center line downwards which created a saddle-shape for the ring. Because you have symmetry turned on, the bottom edge will move inward as well.

Before clicking OK, you can now manipulate other aspects of the model as well. While you are in the "Edit Form" mode, you can click on a "point" where edges meet to make a small change, you can click on an edge to make a larger change, or click on a "face" to make the largest change.

If you happened to click OK before you were ready, that is fine; just right-click anywhere on the screen and select Edit Form again and click where you want to start sculpting.

While you are still in the Edit Form mode, click the intersection point between all four faces on the right-hand side of the cube. Drag that point inward as shown in **Figure 12-18**.

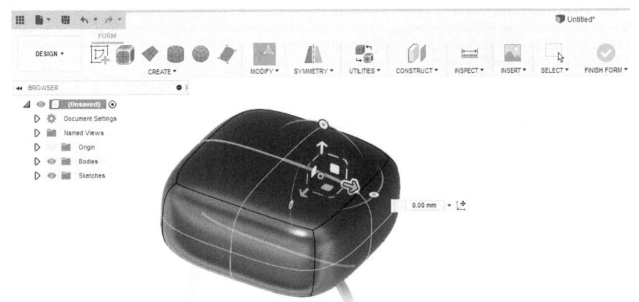

FIGURE 12-16: An edge of the box has been selected, turning it blue. The Edit Form tool has been selected.

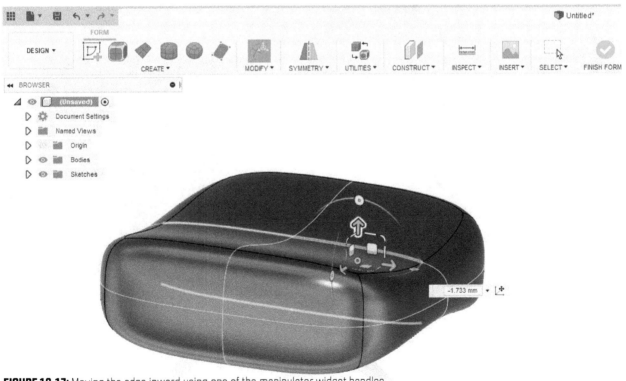

FIGURE 12-17: Moving the edge inward using one of the manipulator widget handles

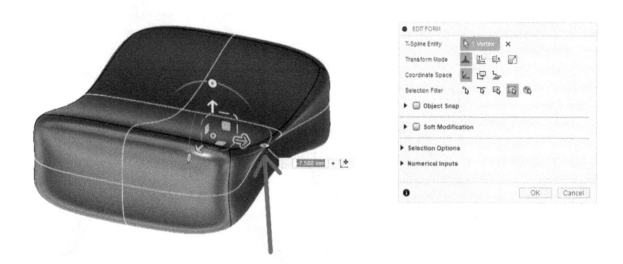

FIGURE 12-18: Within the Edit Form option, a small intersection point at the right was selected and dragged inward

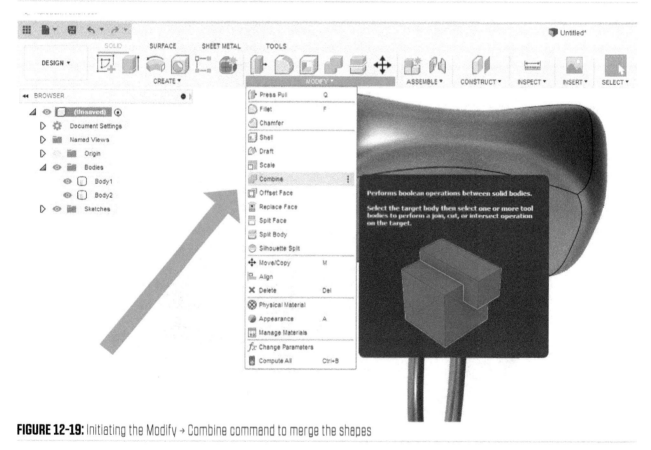

FIGURE 12-19: Initiating the Modify → Combine command to merge the shapes

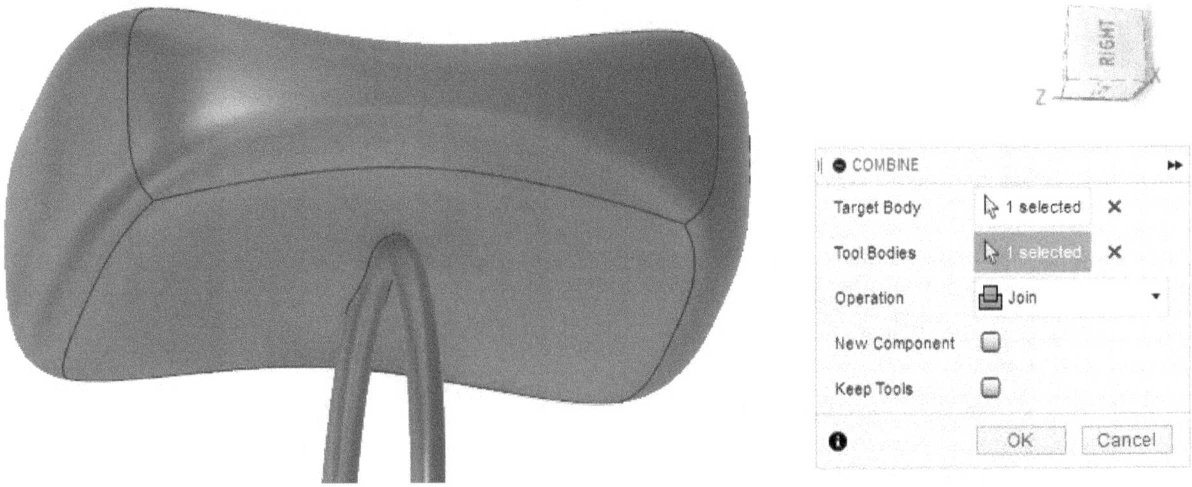

FIGURE 12-20: Target and Tool bodies selected in Combine command

You can also use Edit Form on faces, individual edges, or even the points where the edges meet. Just make sure that whatever shape you create intersects with the band of the ring that you made earlier.

Make sure your Form touches/intersects with the ring! When you have a shape that you like, click Finish Form on the top menu bar to exit out of the sculpting environment. You now have a ring and a topper!

EXPORTING YOUR MODEL FOR 3D PRINTING

The final step is to export the model into your 3D printing software, and you're all set to go.

Fusion 360 can only export one model at a time, so you will need to combine the ring and its topper into one shape before exporting so that it all prints together. If your "topper" does not touch the ring, then you can right-click on the freeform shape, and select "move" and "bodies" in the newly opened window to move it downwards until it intersects the ring. You might have to look from several viewpoints to really see how the two objects are/are not intersecting. The more proper way to do this would be to edit the "Form" you just created by double clicking on it in the timeline and making changes there.

Click the "Combine" option above the "modify" menu as shown in **Figure 12-19**.

Click the ring first and then the topper. One will be selected as the "Target" body and one will be the "Tool" body. Again, make sure that the Combine option is selected as shown in **Figure 12-20**.

Click OK. You now have one shape you can export for 3D printing!

Under the "Bodies" hierarchy in the left menu, "twirl" the triangle next to Bodies, and right-click on "Body 1"

Select the "Save As STL" option and select "High quality" as shown in **Figure 12-21**.

Congratulations! You are all set to load the model into your 3D printer's slicer software, generate support structures, and print a 3D object.

In this chapter you learned how to create a ring with a specific diameter, attach a custom sculpted topper, and combine them together to make a finished model that is ready to export for 3D printing. If you created a version of your own that you really like, post it on Thingiverse.com (or another online 3D model repository) and share it with the world! In the next chapter, we will discuss getting and fixing models to make them print ready!

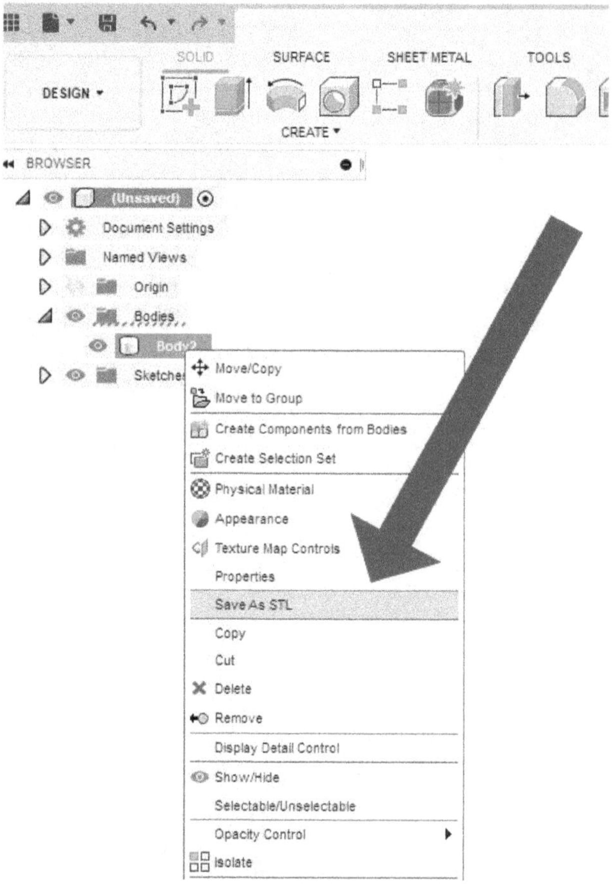

FIGURE 12-21: Selecting Save As STL on the combined body

One More Thing... Combining Shapes from Meshmixer

Remember Meshmixer from the previous chapter? You can bring the ring model into Meshmixer by importing the STL file, and then you can add organic shapes as shown in **Figure 12-22**.

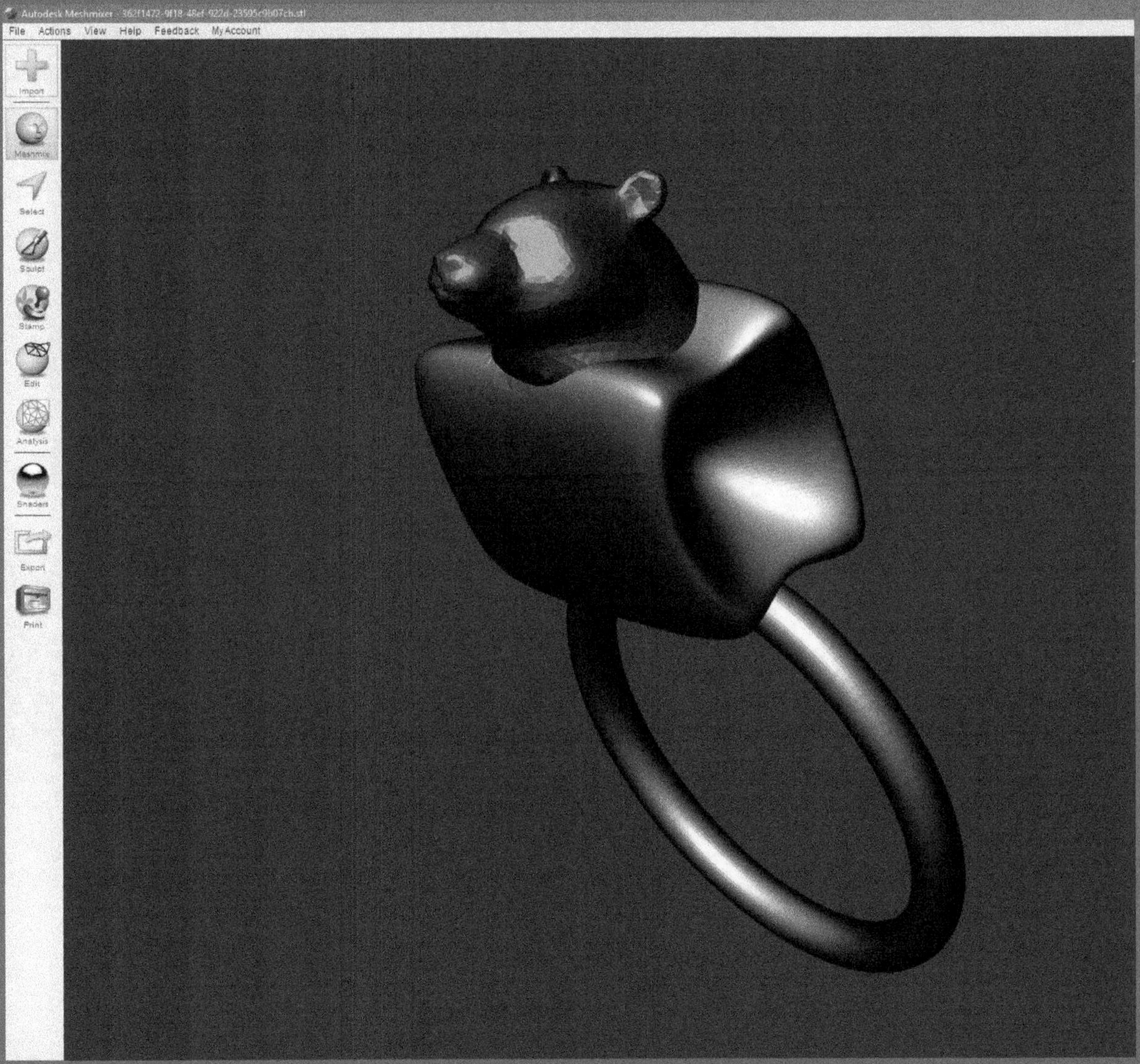

FIGURE 12-22: Your ring with topper imported into Meshmixer with a premade bear's head shape added.

13

GETTING AND FIXING 3D MODELS

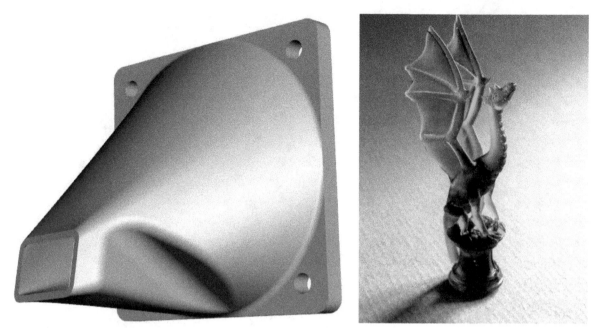

FIGURE 13-1 (LEFT): A 3D model of a fan shroud provided as a suggested first print from the manufacturer of the 3D printer
FIGURE 13-2 (RIGHT): Aria the Dragon (courtesy of Louise Driggers, used with permission)

In the previous chapters we demonstrated how to make your own 3D models, but if you don't want to invest the time learning how to use those programs, you can find a model that's already been created! In this chapter we'll look at the various ways you can find printable models and examine some of the issues you may find with them.

DOWNLOADING A 3D MODEL

For most people who are getting started with 3D printing, it all begins with downloading an existing 3D model. In fact, if you purchase a 3D printer as a kit, there is a good chance that there will be recently-released improvements to your 3D printer you can print. Here is a benefit to using a product that can effectively manufacture parts for itself! The model shown in **Figure 13-1** is a fan shroud, which one printer manufacturer asks you to print for your 3D printer that would otherwise come fully assembled.

There are quite a few sites on the Internet that host free models, but one of the largest sites is called Thingiverse (**thingiverse.com**). Thingiverse is a great place to find 3D models to download and print out yourself because many of the models are free for you to download and print for your own personal use. You can even modify them if you want to, using the software tools we discussed in the previous chapters.

3D MODEL LICENSING AND LEGALITIES

Many of the 3D models you will find for download are provided under what is called a "Creative Commons" (or CC) license (www.creativecommons.org) and they tell you how you can use that model. You should familiarize yourself with the licensing terms for any model you download to print. The beautiful dragon shown in

Figure 13-2 is an example of a user-submitted CC 3D model available for download from Thingiverse.

The Creative Commons license for the dragon model is marked by the author as "Attribution - Non-Commercial" and reads: *This license lets others remix, tweak, and build upon your work non-commercially, and although their new works must also acknowledge you and be non-commercial, they don't have to license their derivative works on the same terms.*

The Creative Commons license is very adaptable, and other models may come with different terms. On many places on the Web, including where you download models from online sites, the Creative Commons icons look like those shown in **Figure 13-3**.

License

Aria the Dragon by loubie is licensed under the Creative Commons - Attribution - Non-Commercial license.

FIGURE 13-3: Figure 13-3: Creative Commons license with icons assigned to Aria the Dragon

The creator of that dragon model (Louise Driggers) licensed that image to us specifically for this book, and the Creative Commons framework helps attribute proper credit to an asset.

Technically, any 3D model can be converted into a file for 3D printing. But the word "technically" is used purposefully. 3D models created for movies or video games are often protected by copyrights and trademarks. If you find a 3D model of your favorite character or item from any prominent entertainment company, make sure it's legal to 3D print.

CREATING 3D MODELS WITH YOUR SMARTPHONE OR DIGITAL CAMERA

Not everyone is willing to put in the time it takes to become adept at 3D modeling and rendering software. Fortunately, there is another way to make 3D models with your smartphone or digital camera. The technique is called "photogrammetry" and is defined as the science of using images to create measurements. In this case, you simply take photos of the object you want to model and upload them to a service that will convert them into a 3D model for you.

In the first edition of this book, there were photogrammetry applications that people could download (even to a cell phone!) and create 3D models for free. As of 2020, however, most of those free tools have fallen by the wayside, and not too many free options are left. If you want a free but feature-limited version of more expensive software, you can look to 3DF Zephyr free **(https://www.3dflow.net/3df-zephyr-free/)** which is limited to 50 photos, but does not cost anything to use. One other completely free option is Meshroom from Alicevision, located at: https://alicevision.org/#meshroom

Another alternative is Autodesk's software program called ReCap Pro **(https://www.autodesk.com/products/recap/overview)** but this one has a monthly subscription of about $40 a month. If you are a student, or are part of a non-profit, the software would be free or very low cost. You will want to take into consideration that Autodesk will also charge you "Cloud Credits" for photogrammetry conversion projects (around $9 each), since all of the work is done on the Autodesk servers.

Although both of the above options are from different software interfaces, the process by which you create a model is similar....and very easy! Note the program you will use: If you are using the 3DF Zephyr free version above, you are limited to 50 photos. Autodesk ReCap Pro allows up to 250 photos per project.

FOLLOW THESE GENERAL BEST STEP PRACTICES:

- Make sure you have enough room to take photos of your object from all sides, from the top, and even, beneath if needed. The space you need around each object will vary but you will need to be able to comfortably walk around the object.

- You can use any camera you wish (even a cell phone camera) and make sure to stand far enough back so that the "fisheye" effect that happens when you are very close to an object is minimized. Each camera lens is different so make sure to do some testing.

- Take pictures of the object from all different angles and elevations, overlapping each picture by about 30-50 degrees depending on how many images you can upload.

- Make sure to raise and lower the camera as you take the pictures to get underneath any overhangs, the top of your object, and other surfaces you can see.

- Load the photos either to the cloud service or into the local photogrammetry application and wait for your model to be exported.

- The resulting file can be edited in Meshmixer, or on any STL modification program.

- If you see that you missed some areas, photograph them again and add them to your past group of photos, run the photogrammetry again and test if the resulting file looks better.

There are some special considerations for creating photogrammetry of objects outside. In this environment the light from the sun and clouds changes quickly so you will have to work fast. Here are some other tips:

- Try to avoid sunny and partially cloudy days. To get the best results you will be balancing the light and diffusing the shadows.

- Use a tripod even though handheld cameras are faster. The shaper your images, the higher quality your scan and textures will be.

- Fill the frame with your object. There is a fixed number of pixels in each image so use them wisely by filling the frame with your object! You will get higher resolution scans that way.

- Use a small fixed aperture size. You want your entire object to be as sharp as possible.

- Use Aperture Priority mode often seen as the Av symbol on your camera. This will automatically adjust the shutter speed to get an optimal exposure.

- Use a Shutter Release Cable to get shaper images. This tip is optional but we found it increases our ability to not move the camera when taking a photograph.

- Take notes on what you scanned and what order will help you organize the images.

- Take real world measurements of what you are scanning so that you can scale your scans to the correct relative size. This is a good tip for indoor or outdoor scanning.

- This last tip is easy: Move fast and stay focused! When taking photos outside, you are at the mercy of the changing weather and light. Capturing the images in the smallest amount of time possible gives you the best chance of capturing images you can use.

After you have had some fun trying photogrammetry with the above tips you can research a promising project called OpenScan (https://en.openscan.eu/). This project aims to provide a set of 3D printable components you would be able to match with low cost stepper motors. This will allow you to automate the process of getting images and to tinker with the hardware setup.

PHOTOGRAMMETRY OF ORGANIC SHAPES

Photogrammetry has another advantage in that it makes capturing organic shapes easier. 3D modeling programs can create complex 3D models from scratch, and by using artistic ability, it can help you create organic models. The problem is that it will take a lot of time to create the curved organic surfaces, whereas with photogrammetry, they can be created with a set of well placed images. Unless you are a trained digital artist, with lots of time, photogrammetry helps to solve this issue by translating complexity in the real world into low-to-medium complexity in the digital world.

Now that you have learned about how to create the images, you will probably want to print the object. If you want to 3D print something you created with photogrammetry, you will most likely have to fix the file before it is 3D printable. In this case, "fixing" means finding the holes in the models where information was not collected and "filling them in" to create a solid model. The work that needs to be done is more of the "cleanup" variety than anything that would require a deep knowledge of 3D modeling. We discussed this process earlier learning about Meshmixer in Chapter 11.

EXAMPLES OF WHAT YOU CAN CREATE IN PHOTOGRAMMETRY

Models created in a photogrammetry application can be used as "starter geometry" and as a basis to build upon. Remember, in the 3D world, size does not matter, so you can create a model of something fairly small and make that 3D model look very large when you mix it with other objects. You can challenge people's assumptions about "what goes with what" in terms of size and function.

HERE ARE SOME IDEAS TO HELP SPUR YOUR CREATIVITY:

- Use photogrammetry to create a 3D model of a large seashell, and then 3D print that model as small earrings. Or do the opposite: 3D print a small shell into a large one.

- Create a 3D model of a couple and 3D print them as a wedding cake topper.

- Give an artist 10 pounds of clay and tell them to sculpt an object. Tell them to be as creative and detailed as they want. The resulting model can then be shrunk down and 3D printed in a smaller size to be, for example, a jewelry pendant.

- Capture a 3D model of an apple tree, and then replace all of the apples with 3D models of sleeping cats (that you also captured with photogrammetry). This would probably be an Internet sensation!

One disadvantage of photogrammetry is the low level of detail the process captures, as shown in **Figure 13-4** of a rock. Even with a good DSLR camera the average detail level of a 3D model that comes out of photogrammetry ranges from 5mm to 10mm. Essentially, the features that are smaller than 5mm will just not show up on the resulting model. 5mm doesn't seem like a lot, but if you look at the transition area between someone's nose and their cheek you'll see it's less than

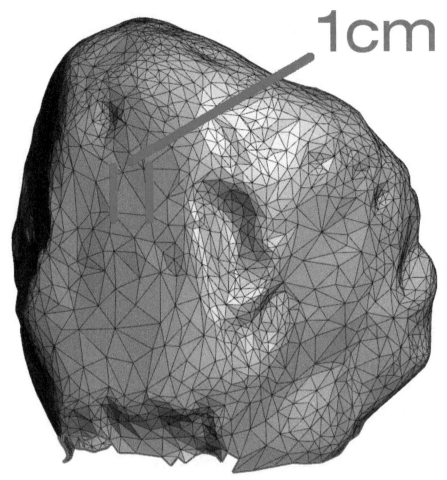

FIGURE 13-4: A garden rock captured as a 3D model using photogrammetry, showing detail and, in some cases, lack of detail. A polygon (triangle) can be as large as 1 cm.

10mm. Even with this disadvantage, photogrammetry's advantage is getting the general shape of an object easily and quickly.

If you are meticulous in the photogrammetry process, you can get excellent results that rival professional scanners in achievable detail levels (50 microns or less), but it takes practice and a controlled environment to achieve those results. The OpenScan project referenced above has some good examples of detail photogrammetry.

As with any technology, there are high-end and low-end devices. In our rapid prototyping division we use two different types of Artec 3D scanners to scan physical objects. They each cost around $20,000 and the phenomenal 3D scans they create (around .1 mm accuracy) can be used for medical, art conservation, and other professional applications. Not everyone has access to a professional 3D scanner, though, and for most purposes a 3D model of average resolution obtained quickly and free of charge provides a great option.

UNDERSTANDING 3D MODEL FILE FORMATS AND UNITS OF MEASURE

The file formats that define a 3D model are universal and can be opened by any number of applications. The major 3D printing file formats in order of popularity from first to last are:

STL (Stereolithographic file)
OBJ (Wavefront OBJ)
AMF (Additive Manufacturing Format, also 3MF)

By a far margin, the most common format for 3D printing is the STL (.stl) file, which describes nothing except the 3D shape itself. It does not provide color, internal object geometry, or really any other information other than the surface shape.

STL CODE

For the coders out there, here is an example of STL file syntax, which can be seen by opening the STL file with a text editor:

```
facet normal 0.000000e+000 -0.000000e+000 -1.000000e+000
 outer loop
 vertex -6.065448e+000 -2.594533e+000 3.400000e+001
 vertex -5.671127e+000 -2.966195e+000 3.400000e+001
 vertex -6.199574e+000 -2.743562e+000 3.400000e+001
 endloop
 endfacet
 facet normal 0.000000e+000 -0.000000e+000 -1.000000e+000
 outer loop
 vertex -6.199574e+000 -2.743562e+000 3.400000e+001
 vertex -5.671127e+000 -2.966195e+000 3.400000e+001
 vertex -6.292432e+000 -2.921260e+000 3.400000e+001
 Endloop
```

The STL file really just describes three sides of a triangle in a 3D coordinate space, and then the next entry (triangle) starts off with one common side from the previous entry (triangle), and defines the next two sides of a new triangle. This repeats over and over again, to create a "polygonal mesh" that all 3D printers know how to process. This file format is very simple and only contains information about the surface or "shell" of the 3D model.

The STL format also cannot contain any color information for printing in color. (We added color to the mesh in **Figure 13-5** to enhance the visibility of the triangles). Nor does that file format have any information about the physical size of the object in the real world. A 3D model in an STL file will show as "x units high," but the physical unit of measurement is not described because size is mostly irrelevant to the modeling program.

In an STL file, something that is 33 millimeters high looks the same as something that is 33 inches high. It is only when you want to print the model that the unit of measurement becomes important.

On a service bureau's website such as **Shapeways.com,** you will be asked to define the unit of measurement as shown in **Figure 13-6**.

CREATING 3D MODELS WITH FOUR-SIDED POLYGONS AND THREE-SIDED POLYGONS

The STL format describes three-sided polygons (triangles) and that is the most common file format for 3D printing. In this book we have also used the term "polygonal mesh" to describe 3D models. This is because meshes do not have to be made of all three-sided polygons.

There are other programs that can use four-, five-, six-, or more-sided polygons to create 3D models. A great example is the open source program Blender (**blender.org**), which has the ability to create multiple-sided polygons called "n-gons" in 3D models. There are benefits to using polygons with more than three sides; it gives the ability to create more graceful transitions between polygons, as opposed to sharp triangular points.

If your model's mesh is made from four- or more-sided polygons, you will need to convert that file into an STL (three-sided triangles) for 3D printing. This conversion from "more than three-sided polygons" down to "three-sided polygons" changes a model by adding in more geometry. This conversion usually works fine, but sometimes creates unexpected artifacts/changes on the new model that were not there

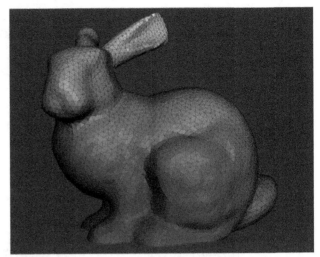

FIGURE 13-5: Visual image of an STL file—a polygonal mesh, or shell, of a 3D model made of many triangles (10,996 triangles to be exact!)

×

Upload your 3D design.

SELECT FILE   No file selected.

Model Units: ● millimeters ○ inches ○ meters

UPLOAD

By clicking "Upload," you are representing that this 3D model does not violate Shapeways' Terms & Conditions and that you own all copyrights for this 3D model or have authorization to upload and use it.

Supported 3D files

Maximum file size: 64 MB or 1 million polygons
Filetypes: DAE, OBJ, STL, X3D, X3DB, X3DV, WRL
For color 3D prints: DAE, WRL, X3D, X3DB, X3DV
Textures files: GIF, JPG, PNG
Upload as ZIP containing model file and textures
Privacy: Private by default

FIGURE 13-6: File upload dialog from Shapeways.com prompting user to define unit of measurement. Note the maximum file size of 1 million polygons. The bunny had just over 10,000 polygons.

FIGURE 13-7: A CAD-modeled rectangular object. Note the perfect rectangles on all sides (also known as "quads").

on the starting model. For example, look at the perfect rectangle in the CAD model shown in **Figure 13-7**.

When you export that model into a polygonal mesh editing program, in order to create an STL file from the CAD file, all of the four-sided polygons (quads) will be converted into three-sided polygons (triangles) as shown in **Figure 13-8**. This happened to work well as triangles can easily make rectangles. In other cases of going from more circular shapes into triangles, the points of the triangles might not translate as well.

FIXING A 3D MODEL FOR 3D PRINTING

All models are not created equal. If you look through a repository like TurboSquid (turbosquid.com), which provides models for game developers, architects, and artists, you will find a lot of 3D models, and many of them are free to download. Since TurboSquid does not focus on models for 3D printing however, many of the models you will find there will not print well. 3D models that are destined for use in computer animation or in computer games do not have to be as complete as 3D models that will be 3D printed. Even 3D models made with high-end 3D scanners and photogrammetry need some post-processing to make them 3D printable.

A common problem you will encounter when you have created a model using photogrammetry or a 3D modeling program is whether or not the model is "manifold." The term manifold refers to how "watertight" the

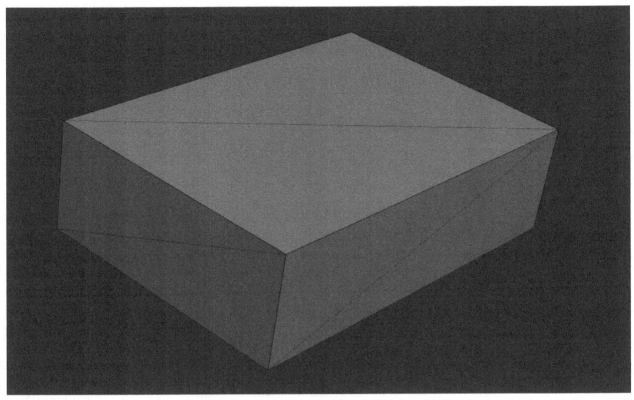

FIGURE 13-8: A rectangular object exported to a three-sided polygonal format (STL). Note that there are no more four-sided areas. They have been converted to triangles.

model is. Think of it this way: all 3D models define a space that is either "outside" or "inside" the model, like the outside or inside of a ball. If you were to fill the inside of a manifold ball with water, *no water would come out*. There would be no holes and no place for the water to leak through.

A non-manifold model is the opposite. Water would leak out of a non-manifold ball because there are holes or missing data in the 3D model. Often non-manifold models have large holes that are easy to see. Sometimes, though, the 3D model may look manifold but really is not. The sphere in **Figure 13-9** looks manifold upon human visual inspection, but the software would find the problem.

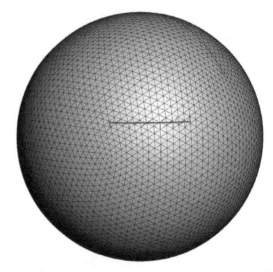

FIGURE 13-9: A small cut (shown in blue) in this 3D model makes the model non-manifold.

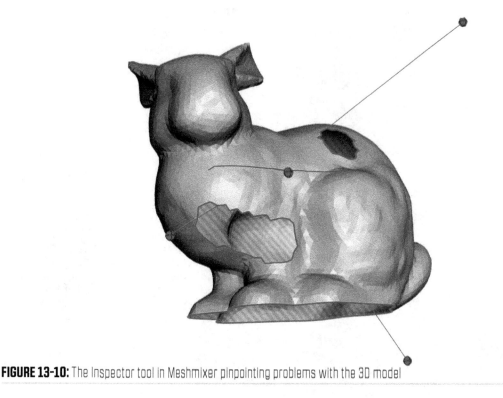

FIGURE 13-10: The Inspector tool in Meshmixer pinpointing problems with the 3D model

USING MESHMIXER TO FIX YOUR 3D MODELS

Without specifically using analysis tools, no amount of visual scrutiny, even if you zoomed all the way in, would show that there was a small cut in this sphere. The slicer software that would 3D print this model would have to make some decisions on how to slice the object for physical printing.

Luckily there is hope! Most of the online 3D printing service bureaus offer tools to repair your 3D model for you. For the most part, these work well, but it is always better to have a good model in the beginning of the 3D printing process. That way, you are more likely to be happy with the end result.

There are a few applications that can help you repair models if your end goal is 3D printing. We believe a very good option is the program Meshmixer, which we mentioned earlier and described in Chapter 11. The Inspector tool in Meshmixer (click "Analysis" then "Inspector" in Meshmixer) can often fix your model in just a few clicks, as shown in **Figure 13-10**.

Clicking any of the little blue bubbles coming out from the model will usually repair that part of the mesh successfully, making the model ready for 3D printing. This tool is so easy to use, it makes Meshmixer an ideal part of your existing workflow for 3D printing, even if you are already comfortable using other tools.

Meshmixer also occupies a unique position in the 3D printing world. It does an excellent job of bridging the gap between the digital world of 3D modeling and the physical world of 3D printed objects. No other software program has such comprehensive tools to help you get a physical object created. If you have not read the chapter on Meshmixer yet (Chapter 11) it would be well worth your time. This free program will help you get more consistently successful 3D prints.

Thinking about using 3D printing to enhance or start a business? This next section is for you! Any endeavor can benefit from 3D printing and the next section gives a great overview of how.

PART IV: FROM IDEA - TO 3D PRINT - TO PRODUCT

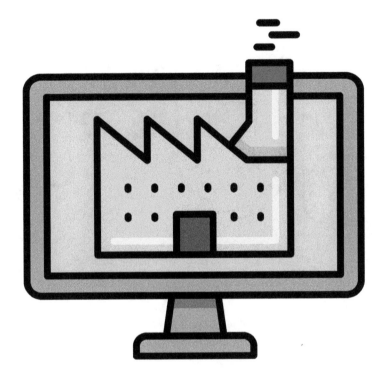

CHAPTER

14

HOW TO START OR ENHANCE A BUSINESS WITH 3D PRINTING

Design Process

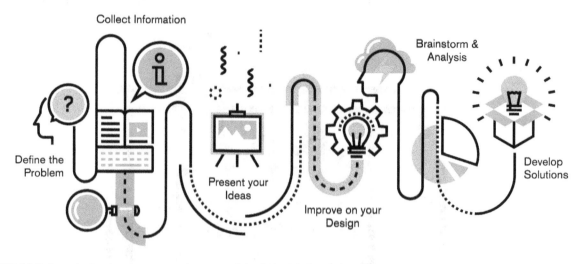

FIGURE 14-1: A graph showing the consecutive steps of the 3D Design Prototyping Process

3D printing is offering entrepreneurs many opportunities to streamline their daily operations and save money. 3D printing also allows for the creation of new parts and models where your imagination is the only limitation.

3D printed items are more than just plastic trinkets. They have real utility and are being used by companies, communities and organizations around the world to change or enhance their organizations. In this next chapter you will learn how you can use 3D printing for your current or new business!

WE WILL EXPLORE TWO DIFFERENT SCENARIOS IN THIS SECTION:

Ⓐ A new company or individual that wants to center their services or products around 3D printing

Ⓑ The existing business that would like to leverage 3D printing to enhance their current sales, marketing and revenue goals

Let's first explore some possibilities available to a new organization that wants to develop a product that can be 3D printed. This type of production method can have many advantages over traditional manufacturing: lower initial capital investment in stock and tooling, an ability to produce small runs quickly and capitalize on trends, and the chance to iterate new designs quickly and relatively inexpensively. We will deep dive into the various production methods in chapter 15. The important takeaway here is that 3D printing allows for more flexibility.

Your new organization has an idea that will change the world, but you must first follow the process of proto-typing and market research. An idea is just the start, to ensure a better outcome, you will want to consider the product design process throughout the prototyping period as shown in **Figure 14-1**.

One you have your CAD model, 3D printing really shows its value by allowing you to create samples quickly and edit them just as quickly. You won't need to wait for weeks for the sample to arrive from overseas. Just load your 3D file into the computer and start the print!

AS YOU DEFINE THE PROBLEM AND COLLECT INFORMATION, YOU WILL WANT TO CONSIDER THE FOLLOWING:

- Do you have any competitors, and if so, how are they making the product or service?
- How much are your competitors charging for the product or service?
- Is your product or service geographically dependent? Does your customer need to be in your area to make the purchase?
- What costs are involved? Do you need to purchase materials? A 3D printer?
- Will you or your staff need special training to run the equipment and what is the learning curve?
- What skills do you bring to the business and what skills will you have to hire?
- How much time and financial resources are you prepared to risk?
- Will you run 3D printers or outsource the production to a 3rd party?
- What 3D printing material is right for your use case? Will it be exposed to heat, light, moisture, etc.?

The above questions are not the complete list of everything you have to consider, but rather a guide to help you narrow down your options. Finally, before you embark on a product or service, we recommend that you familiarize yourself with the ecosystem that makes this revolutionary technology possible. Be sure to research the companies, materials, resources and organizations we listed in Chapter 2 to maximize your success using 3D printing

CREATING PROTOTYPES TO TEST A DESIGN, SHOW POTENTIAL CUSTOMERS, ATTRACT INVESTORS

Prototyping is probably our favorite use of 3D printing and is the main focus of our company. Having a physical prototype to test your invention is critical. There are many advantages to creating that first model: getting mechanical feedback, showing the prototype to focus groups, and getting important market feed-back. Prototyping can also be used to make small volumes to test the market before you invest in large production runs. Small edits to the CAD files can be done quickly and multiple versions can be tested at one time saving time and money.

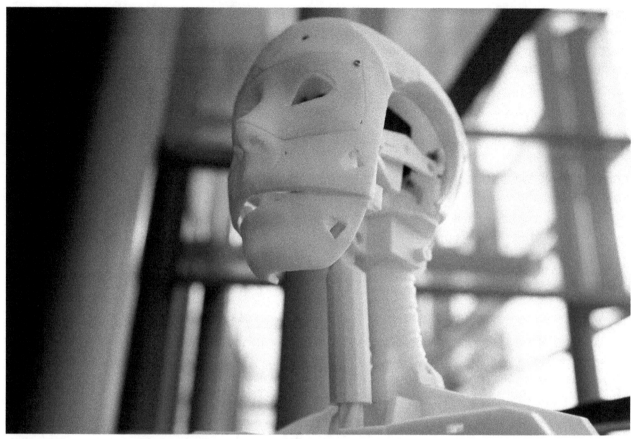

FIGURE 14-2: A humanoid robot prototype developed by testing and iterating 3D printed versions

Trying to raise funds from investors? A photoshopped rendered image will just not do. Investors want to see that the team has done their due diligence by taking the initiative to produce a working model. They want to see that the team has spent time researching and testing the concept. Showing actual CAD models and a history of physical prototypes shows investors that the team has spent time thinking through the design. That reduces the investor's risk and that's a good thing! **Figure 14-2** shows a physical working prototype of a design concept.

STARTING A NEW BUSINESS CENTERED AROUND 3D PRINTING

Social Networks and the Internet have leveled the playing field when it comes to potential exposure to new customers. Entrepreneurs, freelancers and small businesses can start small and build their businesses using a home office. These businesses can reach millions of potential customers with viral campaigns and relatively small amounts of advertising dollars. 3D printing is a perfect tool that can support low volume runs, remote creation, and distribution of goods and services.

And 3D Printing has a broad application. Products that affect your everyday life, like shoes and CPAP masks, as well as highly-complex projects, like buildings and engines, all share a need for customization that can be developed and created using 3D printing. The following examples highlight businesses centered around 3D printing.

YOU CAN CREATE 3D CAD DESIGNS PEOPLE CAN BUY

Did you ever want to create your own jewelry line or model train parts? 3D printing, with CAD knowledge, gives you the freedom to create lines of product your market wants to buy! You can even create the CAD yourself or hire someone to do it for you.

Ready to sell your newly minted creations? Here are some websites where you can you can have a virtual storefront and get discovered by your potential customers:

- Shapeways (https://www.shapeways.com/)
- Etsy (https://www.etsy.com/)
- Turbosquid (https://www.turbosquid.com/)
- Cgtrader (https://www.cgtrader.com/)
- Sketchfab (https://sketchfab.com/)
- Cults3D (http://cults3d.com)
- Prusa Marketplace (https://www.prusaprinters.org/)

FIGURE 14-3: An example of a customized Monopoly game piece being sold to customers

You can create highly specialized products. Whether you like to play video games or Monopoly, you can sell new customized models potential players would love to own. See **Figure 14-3** as an example of a customized Monopoly game piece. There are thousands of games out there with loyal followers and testing a design can be relatively quick and easy. You can jump on fast moving cultural trends and get the benefit of being there first with relevant products to sell.

YOU CAN OFFER 3D PRINTING AS A SERVICE

One of the most obvious ways to make money with 3D printing is to sell 3D prints for a fee. Businesses and individuals often want objects produced through 3D printing, but don't have the equipment. You don't need ten 3D printers to start your own 'Print Farm," you just need at least one 3D printer and the knowledge of how to use the printer. You can buy a ready to use printer, which is slightly more expensive, or you can purchase a good small 3D

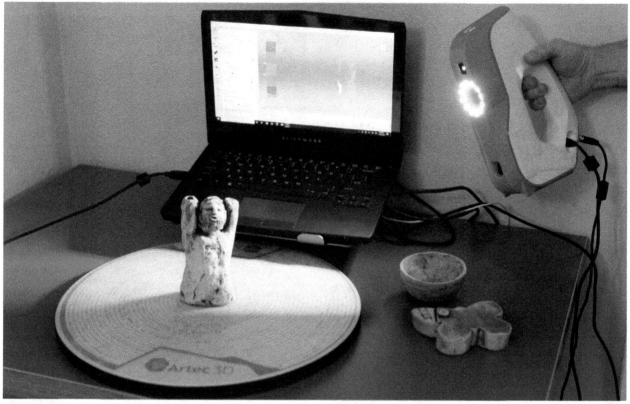

FIGURE 14-4: An Artec 3D Scanner projecting light onto an object in order to create a 3D scan.

printing kit (one you assemble) that can cost around $1,000. Refer to chapters 4-6 for detailed information on the different 3D printing technologies.

Keep in mind you are competing with large service bureaus like Shapeways, Sculpteo and others. Offering great quality and customer service will make your offering more competitive.

HERE ARE A FEW PLACES TO LIST YOUR PRINTING SERVICES SO POTENTIAL CLIENTS CAN FIND YOU:

- 3D Hubs - application required (https://www.3dhubs.com/)
- MakeXYZ (https://www.makexyz.com/)
- Facebook Marketplace (https://www.facebook.com/marketplace)
- Etsy (http://www.etsy.com)

YOU CAN CREATE 3D FILES AS A SERVICE

You need a 3D file in order to 3D print an object. The two most popular file creation methods involve either reverse engineering with 3D scanning (if you have a physical reference part) and CAD modeling (hand mod-

FIGURE 14-5: A trained engineer creating a CAD model by hand using 3D software

eling using a modeling program). These related services are primarily what our company offers and it takes a lot of experience to excel at creating the best models. The learning curve for this business type is much longer than offering 3D printing as a service.

3D Scanning creates a digital copy (scan file) of a physical object, which then allows you to make a 3D print. This is especially important for organic shapes that can't be hand modeled easily. **Figure 14-4** shows a 3D scanner in action, by projecting light onto the object that will later get translated into a 3D map or 3D scan. As we mentioned earlier, you can also use photogrammetry to create 3D printable scans.

3D CAD modeling is more labor intensive and requires a skilled modeler or engineer and access to modeling software as shown in **Figure 14-5**. 3D CAD modeling can be performed by newly skilled workers (for simple shapes) all the way to engineering experts for highly complex models.

Your basic costs will be a computer and the modeling software. This software is quite expensive for the professional versions which can cost $7,000 a year, but many programs, like Meshmixer and Sculptris, are free. There are some powerful programs like Fusion 360 that cost only $600/year (but can be free for some commercial uses if granted by Autodesk), and ZbrushCore used for organic modeling that cost $150 a year. As with all software licensing, you want to make sure you are using the software in a way that is appropriate to your use case. Commercial work needs to be performed under a commercial license, for example.

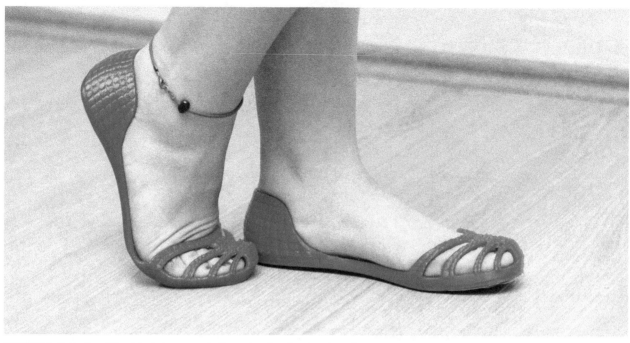

FIGURE 14-6: Custom 3D printed shoes created by using the client's unique foot geometry

3D CAD Modeling is not only the most difficult of the three services mentioned on this list, but also carries some risks associated with this type of artistic freelancing. For example, make certain you agree with your client what services are included in your fee, including the number of revisions. Also, try to establish a working vocabulary with your client, so that everyone agrees with the definitions you are using. And finally, protect yourself legally in case your design injures a third party. Becoming a skilled CAD modeler or engineer takes years and experience. Make sure you understand the liabilities before starting this service.

As you can see it takes very skilled engineers and professional 3D scanning equipment to produce the best files. We invested many years of searching for the best engineers and equipment, but you can start small by taking on easy projects.

WAYS EXISTING BUSINESS CAN LEVERAGE 3D PRINTING
Let's now explore some possibilities available to current companies that want to incorporate 3D printing to increase their margins, garner brand loyalty and create production parts needed to run their equipment.

HERE ARE SOME EXAMPLES:

1. CREATING UNIQUE PERSONALIZED 3D PRINTED GIVEAWAYS FOR CONFERENCE ATTENDEES, AND LIMITED EDITIONS FOR SPECIAL CLIENTS.
Ever go to a conference and get bombarded with conference "swag"? How many pens, mugs and generic

mint boxes can one have! Unique 3D printed items are more exclusive, have a better perceived value and can be customized to that particular conference versus something that is generic and available everywhere.

Company brands can customize certain versions of popular products and offer these limited editions to entice customers to have more brand loyalty. Variations can include: creating a special version based on the date of the event, changing a flagship feature of the model or having the customer's name embedded directly on the design. With 3D printing, changing a feature of the design is relatively easy and fast. Making limited edition parts can happen in days or weeks, instead of months.

3D Printing is ideal for these one-offs and it really makes customers feel special. The reduced cost to produce these small runs can make this option affordable, whereby using a traditional manufacturing method would be too costly and the turnaround time would be too long.

Businesses can also use 3D printing for "Made to Measure" items such as jewelry and shoes using the customer's sizing information as seen in **Figure 14-6**. This will dramatically change the way we shop for everyday items. Ironically, this advanced technology allows for customization like that of the original manu-facturing days of master craft people; making products one by one by hand. Think of the local shoe cobbler but much more efficient.

FIGURE 14-7: An inserted white plastic mold made with 3D printing used in an injection molding machine

FIGURE 14-8: The concept of a gear part as a CAD file, then as a 3D printed replacement part.

2. CREATING 3D PRINTED MOLDS INSTEAD OF COSTLY MANUFACTURING MOLDS

After our company creates the CAD model for a client's design, they then take the files to a manufacturer for a quote if they want injection molding. Part of the cost of injection molding is creating molds to cast the desired design. We will discuss this process in detail in the next chapter, but for now it's an understatement to say that these molds can be expensive. A high quality stainless steel mold can cost $10,000 and up.

The good news is that if you want to produce under 1,000 units, 3D printing may be more cost effective because you don't need a mold. If you do need to go the traditional manufacturing route, you might be able to use 3D printed molds, jigs and fixtures and that can save you a lot of money! **Figure 14-7** shows a 3D printed mold used to manufacture a product.

3. CREATING REPLACEMENT PARTS FOR SPECIALIZED MACHINERY

Custom parts for machinery, especially the ones that are no longer produced, seem to always break down. Before the advent of 3D printing, you would either have to hire an artisan to create a one-off replacement part by hand (for a great deal of money), or make an enormous number of replacement units using traditional fabrication methods, or live without that part, or worse live without the equipment. But that's where 3D printing can really save the day! If you have the original part, a skilled engineer can reverse engineer the physical part into a digital file that can then be 3D printed. Now you can make one or many replacement parts as needed. **Figure 14-8** shows the concept of a machined part that was reverse engineered and replaced with a 3D printed copy.

4. CREATING PHYSICAL PROPS TO GET THE MESSAGE ACROSS.

This is a powerful concept for both internal company meetings and for showing clients what your product can do. If a picture is worth 1,000 words then a physical model is worth even more! With 3D printing the client can see and feel how the product really works.

FIGURE 14-9: Reverse engineered and scaled down version of a jet turbine fan.

One of our clients, a producer of hydroelectric equipment, uses a great "show and tell" prop in their sales meetings. It's not easy bringing a 750 lbs. turbine to meetings, so they wanted us to create something a bit smaller. Through a 3D scanning and reverse engineering process we created a scaled down version that illustrates the physical features. **Figure 14-9** shows an example of a jet turbine scaled down for demonstration purposes.

5. OFFICE FUN- MOTIVATING THE TEAM
If you are in charge of creating motivation at the office then why not try 3D printed items? A 3D printed award or trophy can really motivate the recipient and give them a sense of pride. You can change the trophy label to include their name, date, occasions, and more. All you need is access to someone that can create the CAD model and someone who will 3D print each version.

GENERAL ADVICE
Some of these examples can apply to both the existing business and the new entrepreneur. Our suggestion is to continue to understand the industry's advancements in materials and trends. We also suggest taking inventory of your financial, time and creative constraints when deciding how to use 3D printing for your business venture.

Read the next chapter to gain an understanding of how a prototype becomes a sellable product and learn about more possible manufacturing methods, in addition to 3D printing.

HOW TO MAKE A PROTOTYPE USING 3D PRINTING AND DIFFERENT TYPES OF MANUFACTURING METHODS

FIGURE 15-1: A model of a house made by a model maker using various materials

3D Printing has revolutionized the way we produce physical prototypes, but it's not the only "game in town." To fully understand 3D printing's place in the many trillions of dollars world of manufacturing, you first need to be familiar with the competing technologies that are bigger and have been around for longer. This chapter gives a bird's eye view of the prototyping process, the prototyping process using 3D printing, and the general ways products are made. Suppose you have a conceptual form of your new invention. That's a great start but where do you go from there?

YOU MAKE A PROTOTYPE!

A prototype is an early sample or model of a product, created to test a concept or to act as a thing to be replicated or learned from in the testing phase. It is a term used in a variety of contexts, including design, hardware, software programs. A physical prototype is used to evaluate a new design and make iterations based on the feedback of test users. In this way, prototypes serve to provide a physical validation of a theoretical idea.

THE PROTOTYPING PROCESS

Prototyping is a way of thinking and verifying your idea or concept. We recommend the following method to save you time and to keep you focused.

HERE ARE SOME BASIC STEPS:

- Research your idea: A simple Internet search on Google can give you an idea of what's already in the market and help you build your design.

- Define the features: What does this product accomplish? What problem is it solving? What are the main features?

- Start to make drawings: Whether it's on a piece of paper or in a professional CAD software program, start to draw out the form. In CAD, the measurements will be provided in the model, but in hand drawn sketches you will need to identify key units of measure. Make sure to confirm the design with your partners or investors before proceeding to the next step.

- Create a Prototype: Create a physical representation of your concept with accurate measurements.

- Get feedback: We suggest from a minimum of 10 people. You will document the feedback and modify the design accordingly. This is an essential element of the prototyping process.

- Rinse and Repeat: (Just kidding about the rinse) But seriously, go through the same process mentioned above until you are satisfied with the design and are ready to get manufacturing quotes.

In summary, the rapid prototyping process outlined above offers a quick and relatively inexpensive way to test your idea before spending $10,000 or more on manufacturing tooling. It can help you mitigate manufacturing mistakes, discover important customer feedback, produce multiple design iterations quickly, and generate ideas for your marketing!

GETTING A PHYSICAL PROTOTYPE MADE

In the past, we had fewer ways to make prototypes. For example, model making was a traditional resource as shown in **Figure 15-1**. This option is still available today and materials include, styrene, clay, metal, MDF, acetate, foam board, acrylic, wood, tooling board, plaster, silicone, latex, laminate and fabric. The disadvantages include long lead times, little flexibility for mid project edits, no standard consistency between multiple copies, and a lot of manual labor and cost.

These hand-crafted works of art look appealing, but they rely on the individual skill of each model maker, which can vary greatly, and is difficult to reproduce identical versions. 3D printing takes those variables out of the equation by replicating the model the same way every time (assuming the printers and settings are equivalent). Enterprise level 3D printers even let you prototype in different high-quality

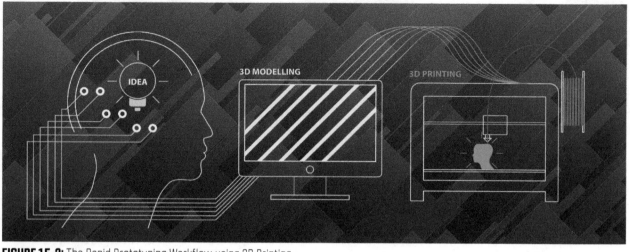

FIGURE 15-2: The Rapid Prototyping Workflow using 3D Printing

materials and colors, all in one run. That's why many people now use 3D printing instead of hiring a model maker.

PROTOTYPING WITH 3D PRINTING

With 3D printing, you can have an idea, create the CAD model and then 3D print the physical prototype as conceptualized in **Figure 15-2**.

Rapid prototyping refers to a group of techniques used to quickly fabricate a scale model of a physical part or assembly using CAD (Computer Aided Design) software. 3D printing can be used to create a physical prototype based on the CAD file. Each new version is created quickly, hence the name Rapid Prototyping. The applications of prototyping with 3D printing differ widely, as workflows and design goals change from industry to industry.

HERE ARE SOME EXAMPLES:

JEWELRY PROTOTYPES: Rapid prototyping transformed not only the design process but also the manufacturing of jewelry. 3D files can be printed in plastic for discussion, testing, research or to get feedback from the client without having to spend money on expensive materials. 3D printing also saves time: going from testing to production in the jewelry industry has become easier especially for the one-of-a-kind pieces as shown in **Figure 15-3**.

ARCHITECTURE PROTOTYPES: Before 3D printing, architects didn't have many options to create their prototypes, so they hired model makers who would carve small scale Styrofoam models by hand in the model shop. Some still use this method, but now architects can also use 3D printing to make models of buildings and structures. Scaled 3D printed models are a cost-effective, precise tool for starting discussions about the artistic, social and functional properties of projects. In addition, the architect can control the reduced

FIGURE 15-3: An example of a prototyped ring, 3D printed in a wax medium, intended to be "burned out" with molten metal via a "lost casting" method.

physical scale much better with 3D printing.

ENGINEERING PROTOTYPES: Engineers, on the other hand, have relied on 3D printed rapid prototyping for decades, as shown in **Figure 15-4**. Accuracy is critical in evaluating a complex design, especially if it has moving parts. 3D printing has the benefit of producing geometrically accurate models quickly 3D printed prototypes are also useful because they can be subjected to rigorous testing and simulation scenarios in CAD programs.

In all industries, many different materials can be used for 3D printing such as plastics, glass filled polyamide, epoxy resins, metals, wax, photopolymers and polycarbonate. This wide variety of options allows users to print their 3D files with materials that have similar properties to the finished product.

MANUFACTURING USING 3D PRINTING

Injection molding, defined later in this chapter, is highly efficient at producing large quantities of a fixed geometry at a low price. However, this cost advantage comes with a striking disadvantage: the inability to produce low volume product lines due to cost. Parts in low quantities have large initial molding investments and may not be amortized over the cost of many individual units. With injection molding, companies are constrained to produce parts and products which conform to the economics, but not necessarily to the demand for variability.

That's where manufacturing with 3D printing can be a better option: You can produce one sample as a prototype or as an end user product. You don't have to wait long lead times to get your prototype back from

FIGURE 15-4: Example of a rapid prototyped new buggy toy model

the factory. You also don't have to wait weeks or months to have expensive molds made. You can own a 3D printer and see the results of your model overnight. If the material and process are right, you can start making products that day!

MANUFACTURING WITH 3D PRINTING HAS OTHER ADVANTAGES LIKE:

- 3D Printing allows for more flexible design choices. You can print more complicated designs that would be impossible to make with traditional manufacturing methods in both plastics and metals.
- 3D printed plastic parts can be an advantage if you need to minimize weight. This is especially important in industries such as automotive and aerospace where light-weighting is beneficial and can deliver greater fuel efficiency.
- 3D printing promotes creativity and innovation as the iterative process is less expensive.
- 3D Printing allows for parts to be quickly created from tailored materials for specific properties such as heat resistance, higher strength or water repellency.
- 3D Printing allows for manufacturing on demand on your schedule.
- 3D printing is more affordable and has less of a start up cost.

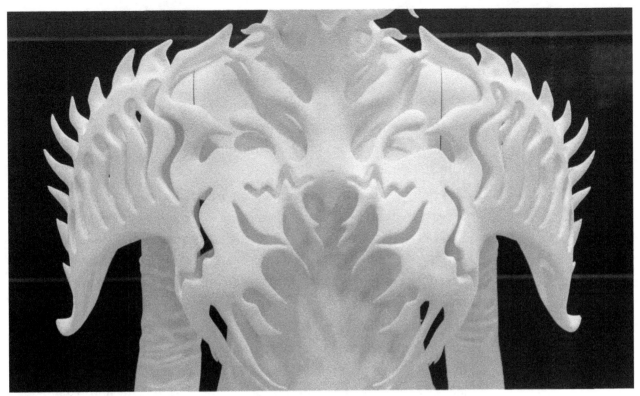

FIGURE 15-5: An artist making the end product using 3D printing.

- 3D printing is more environmentally friendly. It helps eliminate waste by only using the material needed to build a part. Other methods require large blocks of material that the part is cut from.
- 3D Printing saves on resources and storage space and it also reduces the cost of the materials being used.
- 3D printing enables local production. This saves time and doesn't require expensive transportation costs compared to more traditional manufacturing processes produced abroad in other countries.

In general, the economics of using 3D printing for manufacturing work best if you are producing fewer than 1,000 units. **Figure 15-5** shows an example of a product that will probably be made under 100 units total and would, therefore, be more economical to produce with 3D printing.

IT'S IMPORTANT TO NOTE THAT MANUFACTURING WITH 3D PRINTING HAS SOME DRAWBACKS AS WELL:

 3D printing materials are more limited than traditional manufacturing choices (where there are many thousands of options) and so the resulting part might not have the exact same physical and mechanical characteristics as the injection molded option. This is due to the fact that not all metals or plastics can be temperature controlled enough to allow 3D printing

- Many 3D printing materials cannot be recycled and very few are food safe.

- Another disadvantage becomes apparent when you are trying to manufacture many thousands of units. At some point, it becomes more economical to produce the part with other traditional methods than with 3D printing.

- 3D printers currently have small print areas which restrict the size of parts that can be printed. Models that exceed this available print size will need to be printed in separate parts and joined together after production. This can increase costs and time because you would need more print runs and to pay for the manual labor to join the parts together.

- 3D prints need post processing after they come off the build plate. This includes removing support material and smoothing the surface to achieve the desired finish. Post processing methods include water jetting, sanding, a chemical soak and rinse, air or heat drying, assembly and other methods.

- 3D prints are created by adhered layers but these layers can delaminate under certain stresses or orientations. This is particularly more of a concern when using FDM printing.

- 3D printing materials can be more fragile and brittle than parts made by melted and molded manufacturing methods.

- 3D printing may have more measurement discrepancies and looser tolerances. Depending on the printer and material, the printed part might not be as perfectly accurate as the CAD model.

- 3D printing uses more electrical power to produce single units.

- Lastly 3D printing presents a low barrier to entry for copy infringements. An unscrupulous person that also has a 3D printer can copy your product.

Ultimately, you will need to weigh the advantages and disadvantages of using 3D printing as a manufacturing method. You can also decide to start with 3D printing and move to other methods as needed.

Traditional fabrication that has been around for many decades. The journey with these methods can be fraught with obstacles and mistakes can be expensive! They can also be economical and augment your use of 3D printing. The next section of this chapter offers examples of what you might encounter if you select something other than 3D printing.

EXAMPLES OF TRADITIONAL MANUFACTURING

There are many types of processes a manufacturer can use. We will highlight the major high volume traditional production methods. These traditional methods can be grouped into these main categories: casting, molding, machining, joining, shearing and forming. Sometimes a product will use one or more of the above-mentioned categories. With any of these options, the main advantage is lower cost per unit since you are producing many thousands of units at a time.

MOLDING

If the product you're creating starts out as a liquid, chances are the manufacturer will use molding which is the "heavyweight" of the manufacturing world. One popular type of molding involves heating plastic until it becomes liquid, then pouring it into a mold. Once the plastic cools, the mold is removed, giving you the desired shape.

There are four other types of molding: injection molding, which melts plastic to create 3D objects such as butter tubs and toys; blow molding, used to make piping and milk bottles; compression molding, used for large-scale products like car tires; and rotational molding, used for furniture and shipping drums.

The most popular molding process is called injection molding. In the process of injection molding, a hard tool (part of the mold) is created, usually, made of steel or aluminum. The hard tool has an "A side" and a "B side" and was created using a master pattern, usually with CNC machining (computer numerical control) as shown in **Figure 15-6**. When the halves are put together, they contain a voided space within and that void is injected with plastics ranging in material property, durability, and consistency.

HINT:

When making a CAD model for any type of production method, there are some design considerations your engineer needs to consider. These are often called DFM (Design for Manufacturing) or DfAM (Design for Additive Manufacturing). They are general engineering practices of designing models in such a way that will make them easier to produce. For example, adding draft angles (measures in 1-3 degrees) to the sides of a model will reduce the suction force to make the part easier to release from a mold.

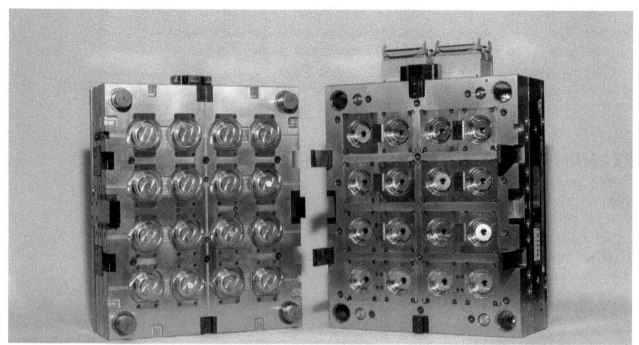

FIGURE 15-6: Injection Mold with an "A" and "B" side for each half of the product

Hard tools created for injection molding are going to be subjected to a lot of stress and heat during the injection process. They will create thousands of parts per day. The care that goes into creating a hard tool involves intense machine programming which costs thousands of dollars alone. The high price for hard tooling is balanced by the mass production capabilities the tooling brings. Plastic cups, dishware, and toys are most commonly made using the process of injection molding because they are common consumer items that need to be produced on a large mass scale.

MACHINING

Manufacturers use tools like saws, sheers and rotating wheels to achieve the desired result when creating something in metal. There are also tools that use heat to shape raw metal material. Laser machines can cut a piece of metal using a high-energy light beam, and plasma torches can cut through hard metals readily using electricity. Erosion machines apply a similar principle using water or electricity, and computer numerical control machines (CNC) use a "subtractive" technology that cuts away material to form the desired form.

JOINING

You can only get so far with molds and machines. At some point you need to be able to put multiple parts together to make one piece. Otherwise, the parts can come apart easily. Joining uses processes like welding and soldering to apply heat to melt together materials. Pieces can also be joined using adhesive bonding or fasteners.

FIGURE 15-7: A Guillotine shear machine is shearing a metal sheet.

SHEARING AND FORMING

Many products start as flat metal sheets. Shearing uses cutting blades to make straight cuts into a piece of metal as shown in **Figure 15-7**. Also known as die cutting, you'll often see shearing used on aluminum, brass, bronze and stainless steel. Another metal-shaping process is forming, which uses compression or another type of stress to move materials into a desired shape. Although forming is often used with metal, it also can be used on other materials, including plastic.

EXAMPLES OF TRADITIONAL LOW VOLUME MANUFACTURING

There are many advantages to using low volume manufacturing that include: Producing a bridge run (more than a few prototypes but less than full-scale production), reducing capital costs by having lower minimums, lower financial risk exposure in case the product needs to be changed after a small batch is created, and having shorter production lead times. This bridge run allows you to get to market more quickly, while allowing the traditional manufacturing's slow ramp-up to happen in the background.

URETHANE CASTINGS (CAST URETHANES)

It's similar to injection molding in that polyurethanes are injected into a tool. But with cast urethanes, the tool is a soft tool, typically made with a type of silicone mold as shown in **Figure 15-8**. Silicone is much cheaper than steel or aluminum but the drawback to using silicone is that it doesn't last as long as injection molding molds. With urethane casting, the mold is created via a master pattern using 3D printing. Cast urethanes are suited for low volume production and prototyping. Because the cost for soft tooling is lower, cast urethanes are an excellent choice for creators still testing product design.

SAND CASTING (SAND MOLDED CASTING)

This metal casting process is characterized by using sand as the mold material. The term "sand casting" can also refer to an object produced via the sand-casting process, which dates back many centuries. Over 60% of all metal castings are produced by the sand casting process and molds made of sand are relatively inexpensive.

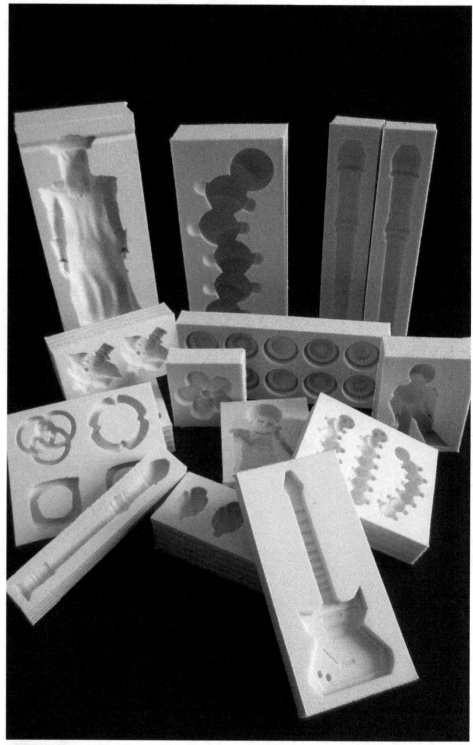

FIGURE15-8: Sets of Handmade Silicone Rubber Molds used in urethane casting

FIGURE 15-9: A foundry worker pours molten metal into sand based casts.

Sand castings are produced in specialized factories called foundries, as shown in **Figure 15-9**. In sand casting, a suitable bonding agent (usually clay) is mixed in with the sand. The mixture is moistened, typically with water, but sometimes with other substances, to develop the strength and plasticity of the clay and to make the aggregate suitable for molding. The sand is typically contained in a system of frames or mold boxes.

CNC MACHINING

As we mentioned above, CNC (computer numerical control) is used to make hard tooling for injection molding. This process starts with a block of material (such as wood or metal), and removes material with drills, boring tools and lathes, controlled by a computer program that is designed to make that pattern. The CNC system transforms a block of material into the precise desired model as shown in **Figure 15-10**. CNC is a subtractive process which removes material, while 3D printing is an additive process that builds material up to create the model.

CNC was a major advancement and improvement over non-computer machining by hand. In modern CNC systems, the design of a mechanical part and its manufacturing program is highly automated. The process is fascinating to watch in person; bits of materials are flying everywhere within the enclosed machines and the precision of the remaining design is impressive.

FIGURE 15-10: Industrial high precision CNC machine creating an object.

CONCLUSION

Bringing a product to market can be exciting, but also expensive and exhausting. Regardless of the manufacturing method you choose, there will be delays, mistakes and process interruptions. The more you research the process, the better you will be prepared to make timely decisions, and the better your results will be. It's impossible to know everything in the beginning and you should expect a learning curve. Hopefully this chapter gave you a glimpse of how to go from prototype to production and how to start a conversation with a manufacturer once you are ready for production.

In the last chapter of this book we will demonstrate how companies are using 3D printing to make prototypes and products. Keep reading to learn how 3D printing will affect us in the future and how it will change our world.

16

HOW 3D PRINTING WILL CHANGE YOUR (AND EVERYONE ELSE'S) LIFE

FIGURE 16-1: An example of a 3D printed torque wrench that helps tighten the Olsson Ruby nozzle to a specific tightness. The manufacturer supplied just the CAD file for this upgrade process.

In this book, you have learned how 3D printers work, how to use CAD software programs, how to set up your own 3D printer, how to use 3D printing in your business and even how to make a prototype! You have also learned about some of the third-party services you can hire to outsource part or all of your 3D printing workflow. We don't have a crystal ball, but this chapter gives a glimpse into what you might expect to see in the years to come.

WE ARE ALL MAKERS, AND COMPANIES WILL FOSTER THAT EVEN MORE

Corporations will help advance the use of 3D printing by encouraging our creativity and enabling on-demand access to products because 3D printing will eventually touch everyone's lives, and companies are starting to notice. Consumers will have a chance to become co-creators as companies actively develop products, services, and tools to help them participate even more in this technology.

PROVIDING ON-DEMAND ACCESS TO PRODUCTS

Hardware stores in the future may not stock all items but instead maintain CAD files of items (even ones that are no longer in production) enabling you to 3D print them at home or pick them up at a 3D print store. If your dishwasher breaks down, the manufacturer may allow you to download the CAD file (for free or for a price) so you can print the replacement part at home allowing on-demand manufacturing...even at 3 am when stores and repair services are usually closed! This saves you the lengthy wait of getting the part delivered in the mail or you having to pick it up at the store. This solution also helps companies from making too many, too few, or from stocking the wrong replacement parts.

Companies that offer these downloadable CAD files as an option might also provide consumer installa-

FIGURE 16-2: In the future, user-friendly interfaces will allow for mass customization of 3D printed goods over the Internet

tion instructions to limit unnecessary service calls. If the part is easy to install, consumers will be able to complete the repair themselves because companies know you want to save time and money. For example, **Figure 16-1** shows a 3D printed torque wrench that is designed for a specific use: to install a 3D print nozzle on a 3D printer with a specific tightness as defined by the internal "fins" on the 3D print. This wrench is not shipped with the nozzle, the end user prints this themselves to assist with the installation process. Everyone wins: The manufacturer doesn't have to incur the cost of manufacturing the wrench, and the consumer doesn't have to purchase it separately.

ORGANIZATIONS WILL USE 3D PRINTING TO STRENGTHEN THEIR CONNECTION WITH YOU

3D printing offers organizations and companies unique branding opportunities. Mass customization through 3D printing allows for the widespread and relatively inexpensive manufacturing of one-off objects. Brand managers know that most customers want products tailored to them. The ability to create something personal, in partnership with their favorite brand, will strengthen brand loyalty and make customers more involved than just passive buyers.

In the future you will see more examples of branded "one-off" gloves, hearing aids, hats, and more. Through intuitive software, you might be able to adjust the design of the product you are buying. It will be the new way to monogram your purchases, one can only imagine that someday it might all be possible with just the touch of a button, as shown in **Figure 16-2**.

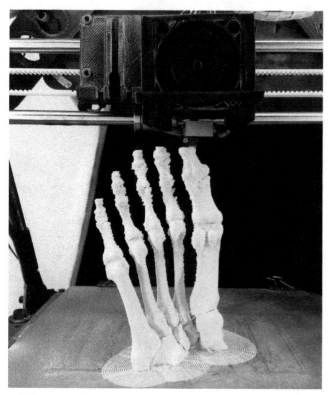

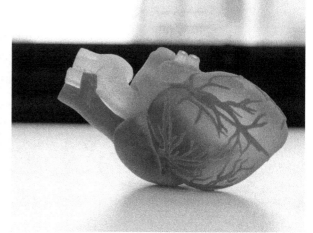

FIGURE 16-4: A model of a 3D printed heart used in medical training. Note that the inside vascular system is hollow, allowing for procedures to pass through the chambers.

FIGURE 16-3: A medical school lesson includes a 3D printable file as part of the curriculum. Shown here is a 3D printed foot chosen by the textbook author for use in an anatomy class.

TAILOR MADE FOR ONE

Companies are motivated to acknowledge your unique needs, interests, and tastes because doing so strengthens their connection with you and increases sales. They will take your design aesthetic and/or actual physical measurements into consideration when they make your (literally your) item. Therefore, competitors that don't offer you a 3D printed customized option might come across as uncaring or unwilling to acknowledge your preferences...and that never goes well for sales.

But in order for this to occur, companies will need to provide a very easy-to-understand user interface when presenting their customers with so much choice to minimize confusion. This combination of challenge and opportunity will change the way companies offer choices and promotional items to their customers.

3D PRINTED EDUCATIONAL TOOLS

Educational textbook companies have already started to adapt this technology into their curricula. Students can view 3D models online provided by the authors and download the files to create 3D printed learning tools, as shown in **Figure 16-3** and **Figure 16-4**.

ORGANIZATIONS WILL INCREASE PROFITABILITY AND BE MORE ECO-FRIENDLY

3D printing will enable all organizations (both for profit and nonprofit) to benefit from more than just brand loyalty and increased sales. 3D printing will allow organizations to increase their profitability and decrease their environmental footprint. Organizations will be able to make just what they need, as they need it, locally, without the waste normally associated with mass production. Additionally, if a design flaw is found in a product, instant changes can be made at a much lower cost than would be the case for retooling the entire production line. Not to mention, not having to throw out thousands of faulty or outdated products that would normally go into landfill.

FIGURE 16-5: Excess inventory that may be defective, not needed or no longer in style

3D printing can drastically change the supply chain by bringing manufacturing closer to the end user and initiating the manufacturing when a customer wants to buy it, not before. This is called "just in time" manufacturing. Environmentally conscious consumers will rejoice in 3D printing's ability to produce one-offs that will help control inventory waste, reduce shipping distances, and shorten lead times. 3D printing will also help reduce these kinds of vast storage needs, as shown in **Figure 16-5**.

Organizations that stock more CAD files than physical items can potentially reduce the amount of wasted inventory and, therefore, become more ecologically responsible. The distinction between shopping online and shopping at a brick-and-mortar location may become less clear as your local garden supply store, clothing store, or even pharmacy begins to include CAD files in its inventory alongside the physical items on the shelves.

LOCAL ECONOMIES WILL BENEFIT FROM THE COMMERCIAL USE OF 3D PRINTING

As mentioned before, 3D printing will not replace mass production but rather augment current manufacturing methods. Specialized makers, potentially including you, will have the ability to economically manufacture "low volume" production runs ranging in numbers from 1 to 1,000 units.

With 3D printing, you will see products being made for hyper-local markets like cities, towns, or even neighborhoods. This hybrid model will help keep more revenue sales local. In addition, local economies will also get the tax revenues associated with the sales transactions.

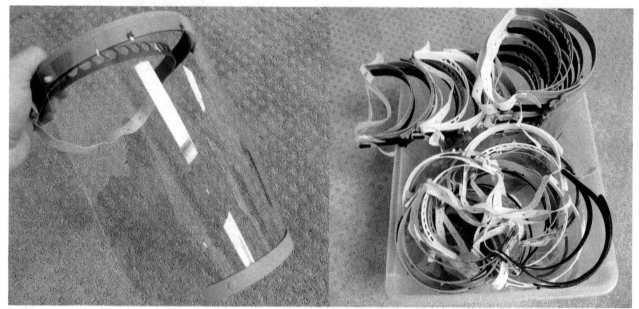

FIGURE 16-6: Open source face shields 3D printed and assembled by HoneyPoint3D on our Prusa Printers (https://www.prusa3d.com/covid19/) and sent for donation to the UCSF hospital.

There are organizations around the world that want to encourage an environment of local manufacturing. One such organization is America Makes (**http://www.americamakes.us**), a premier national accelerator for the 3D printing industry that encourages the growth of US-based manufacturing. A private/public initiative formed in 2012, America Makes aims to bring back local jobs and manufacturing to the USA through the application of Advanced Additive Manufacturing (another name for 3D printing).

But job candidates will have to adapt. Local job seekers, wanting local manufacturing jobs, will have to acquire new skills that prepare them to work in this new manufacturing environment. Candidates entering this workforce will have more opportunities if they are proficient in CAD modeling, 3D printing management, and the related technologies.

MEDICAL APPLICATIONS OF 3D PRINTING

This next section highlights the affects 3D printing has and will have in the medical community, perhaps one of the biggest early adopters of this technology.

COVID-19

We couldn't write this book without mentioning the contribution 3D printing has made in the fight against the Covid-19 pandemic. In the beginning of 2020, Covid-19 had negative repercussions in every industry. One of the earliest industries to be the most vulnerable was the medical sector. This industry was caught by surprise and the need for both PPE (personal protective equipment) and other medical equipment exceeded the supply on hand.

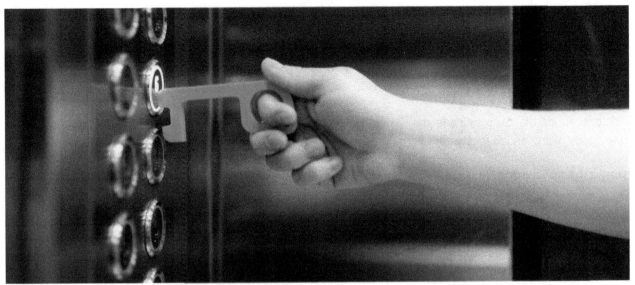

FIGURE 16-7: A door opener and button pusher developed using 3D printing technology in order to keep the user germ free.

When the government and hospitals ran out of these supplies, the 3D printing industry rushed in and started prototyping and making everything from nasal swabs to ventilator machines parts. The speed and sense of urgency felt throughout our industry was palpable, inspiring, and humbling. Many thousands of engineers, companies and hobbyists donated their time to make anything that was in shortage. Soon 3D printed face shields, ventilators, and later, test swabs were available to those that desperately needed them. **Figure 16-6** shows the face shields we made and donated to the UCSF medical hospital. That donation was a fraction of what they needed but they were still appreciated.

FIGURE16-8: 3D printed replacement parts that were critically needed to run medical equipment.

Now more than ever, the importance of 3D print-ing and on-demand manufacturing can't be ignored as critical gaps were filled when traditional supply chains fell short. As the 3D printing industry offered, many times at no cost, to develop, make and distribute critical supplies that were not being met by the traditional supply chains, the world saw the value in local manufacturing at speed that met the demand.

We don't know and may never know the extent of how many lives were saved because of 3D printed PPE. However, with certainty, it was an incredibly proud moment to see our industry mobilize so quickly and show how 3D printing could be useful in widespread emergencies. **Figures 16-7** and **16-8** show more examples of 3D printed items that were developed and printed to fill the supply chain gaps.

The perception that 3D printing is just for toy trinkets was officially over. The new era of 3D printing and its use cases has changed forever in the minds of consumers and with other industries alike.

PRESURGERY TOOLS

The images on the next page show how 3D printing can change pre and postoperative medical treatment. If a surgeon can 3D print the actual anatomy of the patient before surgery, the doctor can better plan the procedure and even consult with other doctors to increase the chances of better outcomes. With 3D printing, the doctor has the time to think through the surgical plan before the patient is on the operating table. **Figure 16-9** shows a patient's blood vessel that was 3D printed for pre surgery analysis.

CUSTOMIZED MEDICINE

All societies and consumers enjoy the benefits of customization. They want custom made tables, shoes, pools, houses and basically anything! With so many advancements in modern medicine, it wouldn't be surprising to see in the future 3D printed treatments tailored to each individual. Some people might need more or less of an active ingredient, and therefore, 3D printed pills for each individual may be an option some day as shown in **Figure 16-10**. Imagine having custom made Aspirin or Tylenol with doses that are made for your genetics!

If you think personalized medicine is interesting, how about 3D Bioprinting? What if we could 3D print organs and lessen the burden of organ shortages? This may become a reality as the European consortium OrganTrans **(https://organtrans.eu/)** is currently trying to prepare and develop a tissue engineering platform capable of generating liver tissue. This proposed automated and standardized alternative solution to organ donation for patients with liver disease could be revolutionary.

3D PRINTING FUTURE TRENDS AND PREDICTIONS

As this technology continues to grow rapidly, here are some more future trends you can look forward to experiencing in the years ahead:

- 3D Printing will be used for mainstream production as higher volume production becomes more financially viable.
- Design software will become more integrated and easier to use. In order to 3D print, you need a 3D file. When the software becomes more intuitive and easier to use, higher adoption will follow.
- A broader educational reach in schools will increase the available workforce and further all industries adoption.

FIGURE 16-9: An example of a medically accurate human blood vessel, 3D printed in resin

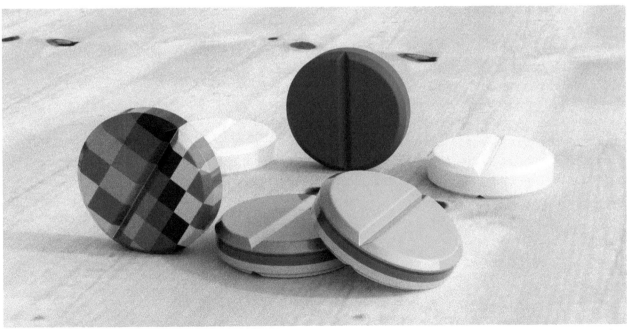

FIGURE 16-10: Illustrates the concept of highly personalized 3D printed medicine. Each layer can be printed with specific amounts of the active ingredients.

- 3D printing will become a central tool in dentistry. Currently, dental crowns, surgical guides and dental aligners all use 3D printing and the applications will continue to expand.

- 3D printing will become smarter with sensors and machine learning systems. Sensors and cameras will provide real time feedback and affect the in-progress printing.

- Metal 3D printing will continue to improve and quality assurance will increase, making this material more applicable to manufacturing.

- Automation of the post processing part of 3D printing will increase the profitability of each unit. This will enable the technology to compete more with traditional manufacturing methods.

- Material options and advancements will continue to increase. This expansion of printing materials will attract more industries and use cases. For example, composite materials that are lightweight and strong, can help reduce manufacturing and energy costs.

GET INVOLVED AND GET CONNECTED. YOU ARE PART OF THIS FUTURE!

The learning doesn't stop here. You are a homesteader of this advancing technology and the field will continue to grow and change rapidly in the coming years.

For that reason, we encourage you to engage with the 3D printing community. There are many member-driven groups you can join. One of the largest online groups for 3D printing is the LinkedIn community for 3D printing, with around 63K members. **Figure 16-11** is a screen capture from their webpage. Another popular community is on Facebook as shown in **Figure 16-12**.

If you are a woman or want to support women in 3D printing, there is an organization called "Women in 3D Printing" that has over 75 world wide chapters, a magazine, events and currently has over 17,000 members. **https://womenin3dprinting.com/**

There are other ways to connect with the 3D Printing community. We recommend that you find online forums that suit your level and interests, and upload your 3D model creation to online repositories to get feedback from the community **(https://reprap.org/wiki/Printable_part_sources)**. You could also attend live 3D printing events, Meet-ups and webinars in your area. It's a great way to stay informed and up-to-date.

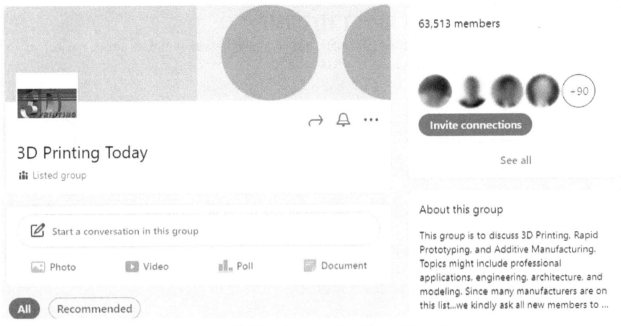

63,513 members

Invite connections

See all

3D Printing Today

Listed group

Start a conversation in this group

Photo   Video   Poll   Document

All   Recommended

About this group

This group is to discuss 3D Printing, Rapid Prototyping, and Additive Manufacturing. Topics might include professional applications, engineering, architecture, and modeling. Since many manufacturers are on this list...we kindly ask all new members to ...

FIGURE 16-11: The main landing page of the LinkedIn 3D printing community with over 63k members

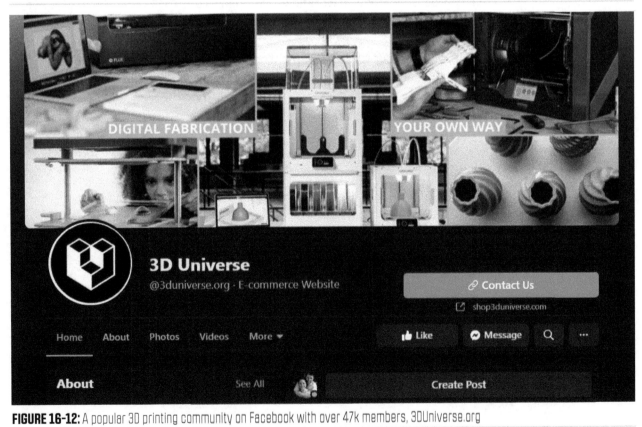

FIGURE 16-12: A popular 3D printing community on Facebook with over 47k members, 3DUniverse.org

THE BEST IS YET TO COME

We already know, 3D printing has changed the way we create and use objects. In the future, 3D printing will change our lives financially, psychologically, socially, and creatively.

3D printing will change how we make everything from personalized pharmaceutical pills to how we build custom houses. 3D Printing will continue to change our lives in ways we can't even imagine.

With 3D printing, we will no longer be bound to the set list of possible choices offered by a manufacturer. We become the manufacturer ourselves.

To make this a reality, the industry and technologies will continue to advance as more content is created, industry standards are better defined, and the process of creating 3D models is made easier.

3D printing is still relatively young in its application and it is expanding every year with new advancements in workflows, companies and materials. The road is not 100% certain and not 100% clear except that 3D printing will continue to evolve.

As you continue your journey into the world of 3D printing, keep in mind that the most important application may have yet to be discovered. Maybe it will even be created by you!

Photo courtesy: HoneyPoint3D™

LIZA WALLACH KLOSKI CO-FOUNDER, HONEYPOINT3D™

Liza Wallach Kloski was born in Guadalajara, Mexico and founded Liza Sonia Designs in 2003, a unique upscale jewelry brand and retail store in Northern California, which wholesaled jewelry to over 100 retail stores including 17 Nordstrom department stores. A graduate of UC Berkeley, Liza has won numerous design and business awards and was the main retail expert in Entrepreneur Magazine's paperback book "Start Your Own Fashion Accessories Business (StartUp Series)." Liza co-founded HoneyPoint3D and now heads Sales and Operations.

NICK KLOSKI CO-FOUNDER, HONEYPOINT3D™

Nick has 16+ years in the high-tech industry. Graduating from UC Santa Barbara with an English Major, he was hired by Sun Microsystems during the dotcom boom, and has held a number of technical roles at Sun, and Oracle, translating complex technical architectures into understandable ideas. Nick's skills go deep into both the technical understanding of the industry and the mechanics of 3D printing, CAD and 3D scanning. Nick is an Expert Elite member at Autodesk and won an award from Autodesk for his Fusion360 course. Nick co-founded HoneyPoint3D and now heads the Education and 3D Services Division.

Nick and Liza are considered 3D printing experts and have been interviewed on CNN, RT Television, San Francisco Business Times, KGO Radio, other Bay Area newspapers and have been speakers at many industry conferences.

They are the co-authors of Make Magazine's 1st edition book on 3D printing titled, "MAKE: Getting Started with 3D Printing," that was released in May 2016. The book has over 135 "FIVE STAR" reviews and was translated in Mandarin. The second edition of this book was published in 2021.

ABOUT HONEYPOINT3D™

HoneyPoint3D was founded in 2012 by husband and wife team, Nick and Liza, and is now one of Northern California's premier engineering companies offering Rapid Prototyping and Development, 3D Modeling, 3D Scanning and 3D Printing services to private and publicly traded companies. HoneyPoint3D licenses online courses to LinkedIn and also has a national contract with The UPS Store offering services to centers that participate in the program. Visit www.HoneyPoint3D.com for a press kit.